WORKBOOK for

MACHINE TOOL PRACTICES

Eighth Edition

Richard R. Kibbe
John E. Neely
Roland O. Meyer
Warren T. White

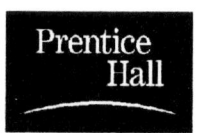

Upper Saddle River, New Jersey
Columbus, Ohio

Production Editor: Christine Buckendahl
Design Coordinator: Diane Ernsberger
Cover Designer: Jeff Vanik
Production Manager: Deidra Schwartz
Marketing Manager: Mark Marsden

Pearson Education Ltd.
Pearson Education Australia Pty. Limited
Pearson Education Singapore Pte. Ltd.
Pearson Education North Asia Ltd.
Pearson Education Canada, Ltd.
Pearson Educación de Mexico, S.A. de C.V.
Pearson Education—Japan
Pearson Education Malaysia Pte. Ltd.
Pearson Education, *Upper Saddle River, New Jersey*

Copyright © 2006 by Pearson Education, Inc., Upper Saddle River, New Jersey 07458. All rights reserved. Printed in the United States of America. This publication is protected by Copyright and permission should be obtained from the publisher prior to any prohibited reproduction, storage in a retrieval system, or transmission in any form or by any means, electronic, mechanical, photocopying, recording, or likewise. For information regarding permission(s), write to: Rights and Permissions Department.

10 9 8 7 6 5 4 3 2

ISBN: 0-13-172103-8

CONTENTS

INTRODUCTION..1

PROJECTS

    1. Drill and Hole Gage..3
    2. Turning and Threading Project............................9
    3. Making a Center Punch.....................................17
    4. C-Clamp...25
    5. Paper Punch..33
    6. Parallel Clamp...43
    7. Tap Handle..51
    8. Die Stock...61
    9. Center and Taper Sleeve...................................71
    10. Internal Threads...77
    11. Multiple Lead Threads......................................81
    12. Acme Threads...87
    13. Carbide Tool Exercise.......................................91
    14. Vee Blocks...97
    15. Precision Vise..111
    16. Cylindrical Grinding...131
    17. Tool and Cutter Grinding................................137
    18. Pin Punch Set..143
    19. CNC Programming Project # 1.......................149
    20. CNC Programming Project # 2.......................155

ADDITIONAL PROJECTS

    1. Wheel Puller...165
    2. Machine Vise..179

TABLES

    1. Decimal and Metric Equivalents of Fractions of an Inch...............197
    2. Tap Drill Sizes for Unified and American Standard Series Screw Threads..198
    3. Metric-Inch Conversion Table.............................200
    4. Inch-Metric Measures...200
    5. Useful Formulas for Areas and Dimensions of Geometric Figures...201
    6. Allowance for Fits of Bores in Inches..................201
    7. Tapers and Corresponding Angles......................202

INTRODUCTION

This workbook is designed to be used in conjunction with <u>Machine Tool Practices.</u> As you encounter each phase of a project or exercise, you will be instructed on the worksheet to study and complete Post-Tests for certain units in the textbook. If you are directed to refer to certain units in the textbook, this means you should use the units merely to refresh your memory for the operation, since you will have already completed the study and Post-Tests for those units in an earlier project. It is absolutely necessary that you follow the routine and sequence of study. Without the related theory, it would be very difficult, if not impossible, and definitely hazardous to you to try to complete the project in this workbook. Do not make the mistake of putting off doing your "paperwork" until after the project is finished, since that would defeat the purpose of the learning system and you would miss some important training. In most cases, credits and grades will not be given on your project until the required Post-Tests are completed and turned in to your instructor. Without your instructor's approval, you should not begin the next project or exercise.

Your textbook contains answers to the Self-Tests in each unit in the appendix. The Self-Tests will help you to check your knowledge of the subject matter before taking the Post-Test. Also, in the textbook appendix, are a number of helpful tables and useful formulas. For your convenience, some useful information and tables have also been included in the appendix of the workbook, since it will always be nearby or at the machine on which you are working.

The project instruction sheet organizes your study in the textbook and in the workbook worksheets. When a unit of study in the textbook has been completed and the Post-Test taken, the worksheet for that unit that is in your workbook should also be completed, unless your teacher has given you instructions to omit that worksheet. When you have completed your project, take it to your instructor for evaluation. Also tear out the project grading sheet which precedes the worksheet and give it to your instructor.

These projects and exercises were planned to give you some training and experience in using machine tools in various ways to produce useful parts and mechanisms. In machine shop work, tolerances are very important. On the beginning projects the tolerances are very broad; standard tolerances (plus or minus 1/64 inch) are used. However, in later projects the tolerances become closer; in some cases, within tenths of one thousandths of an inch.

It is important for you to know how to follow drawings without error to make precision parts, since this is what you must do as a machinist. You will be able to achieve this ability without great difficulty if you follow the procedures given to you in the textbook and this workbook.

Since the main objective of the major lathe project is to give you more experience in using the lathe in a variety of operations, the project must be more complex than any you have completed up to that point. This project will also bring you briefly into some basic milling operations. You must therefore plan to spend a great deal of time completing this project, perhaps as much as 100 to 200 clock hours in the shop.

Major lathe projects are provided in this workbook so that you may have a choice. Your instructor may also have other selections for you to choose from. In any case, your selection must be made with the permission of your instructor and with his or her advice.

For any number of reasons, you may not wish to take the time to do the most time-consuming or challenging project and so may wish to choose the less difficult one. The wheel puller is treated as an alternate project and may be found in the alternate project section.

The precision vise project also has an alternate, the machine vise. This, too, gives a choice between a more exacting project requiring extremely close tolerances and precision grinding techniques and one whose tolerances are broader and which requires only milling machine finishes. Again, the choice must be subject to the advice and supervision of your instructor. Your instructor may or may not assign greater or lesser credits or grades to these projects; you should find out before choosing.

The final term project, if required may be of your own design, planned and drawn to scale. The choice of project will be subject to your instructor's acceptance and permission while it is still in the sketching stage. It must then be drawn to scale with assembly and detail drawings. The device must then be made in the machine shop, according to the drawings. Some examples of these final projects are drill presses, power hacksaws, hydraulic gear pumps, and wood lathes.

It is our hope that the textbook, <u>Machine Tool Practices,</u> and this workbook will help give you a better understanding of the machining trade and make you a more knowledgeable and capable machinist on the job.

PROJECT 1. DRILL and HOLE GAGE

INSTRUCTOR EVALUATION

NAME_____DATE_____

Grade

	Letter	Percent
1. Follows drawing (dimensions and tolerances)	_____	_____
2. Machining finishes	_____	_____
3. Mechanism or tool operates satisfactorily	_____	_____
4. General workmanship	_____	_____
Total grade	_____	_____

Comments

Instructor signature: _____

PROJECT 1. DRILL and HOLE GAGE

Objectives

1. Learn general safety in the machine shop.
2. Lean how to use power cutoff tools and hand cutting tools in bench work.
3. Learn to measure and layout workpieces accurately.
4. Prepare for machining operations.
5. Be able to select drills correctly from any of several series.
6. Be able to offhand sharpen twist drills on a pedestal grinder
7. Be able to drill holes accurately with a drill press.

Outline for Study

Prior to starting each procedure for this project, study and complete the applicable units in the textbooks and the Post Tests for:

1. Safety instructions: **Section A**
2. General shop information: **Section B**
3. Using the steel rules: **Section C**
4. Preparing to use machine tools: **Section F**
5. Cutoff Sawing: **Section G**
6. Layout: **Section E**
7. Using hacksaws and files in bench work, **Section B**
8. Selecting drills and drilling procedures: **Section B**

Procedures

Begin the procedure for this project by completing the following worksheets:

Worksheet 1. Cutoff sawing

Worksheet 2, Layout

Worksheet 3. Making the drill gage.

      A. Sawing the shape of the drill gage

      B. Filing the surface of the drill gage

      C. Drilling the holes

For drill grinding procedures and drill selection, refer to Section H.

WORKSHEET 1. Cutoff Sawing

Exercises

1. Obtain some scrap material and make some practices cuts using the reciprocating power hacksaw and the horizontal band cutoff machine.

2. Cut material for the drill gage project using the reciprocating or band cutoff machine.

WORKSHEET 2. Layout

Exercises

Obtain a piece of scrap material. Properly prepare the material and do the following layout. Alternatively, do, only the layout of the drill gage as shown in the textbook, Section E.

WORKSHEET 3. Making the Drill Gage

A. Sawing the shape of the drill gage

Materials

    Hacksaw
    Vise
    1/8 x 2 cold finish (CF) steel bar stock

PROCEDURE

Sawing

1. The length of the material should be cut 1/16 in. longer than the finish size of the drill gage.

2, Make the angular saw cut #1, leaving 1/32 inch of stock for filing as shown in the drawing. Hold work so that the saw cut is in a vertical position. Do not saw to the end of the line. Leave 1/16 inch.

3. Clamp the material so that the saw cut is vertical. Keep the cut as close top the vise jaws as possible to prevent vibration. Saw as far as possible with the saw blade in the normal position. When the limit of the frame has been reached, turn the blade 90 degrees and continue sawing.

4. Leave 1/32 inch outside the layout line for filing. If the kerf begins to wander in or out, you cannot correct it by angling the saw, cut off the scrap piece from the outside edge to the end of the saw cut and start a new cut.

5. Continue the cut until it joins the angular cut. Be careful and do not undercut a the corner.

Tools and Materials

Selection of files
Bench vise
Previously cut drill gage as shown

PROCEDURE

1. Rough filing on the sawed surface, side #1 may be done by cross filing. If a large amount of stock needs to be removed, use a double cut file. Leave 1/64 inch for finish filing.

2. Using a single cut file, all the straight surfaces may be finished by draw filing side #2. Check frequently with a rule or straightedge to make sure that the surface is flat. The high spots are easily seen and should be filed until almost no light is sees between rule and part.

3. Rough file the radius, side #3. with a double cut fie, leaving 1/32 inch for finishing.

4. With a single cut fie, a shown as side #3. As you file with a forward stroke. move the handle downward.

5. Check frequently with a 1/2 inch radius gage. Continue to file on the high spots until almost no light is seen between the gage and the part.

Tools and Materials

No. 2 center drill
Countersinks
Parallels
Drill press vise
Cutting oil
Fractional inch drills in diameters of: **1/2** (.500), **7/16** (.4375), 3/8 (.375), **5/16** (.3125), **1/4**(.250), **3/16** (.1875), **1/8** (.125), **1/16** (.0625).

PROCEDURE

1. Test all of the drills you are going to se on similar material to your drill; gage. Resharpen as needed.

2. After laying out and center punching for the holes locations, as in the drawing above, set up the drill gage in the vise with parallels arranged under the each edge so that the largest drill can clear the when you drill through.

3. Place the center drill in the check and tighten. With the machine off, spot the center drill in the punch mark of the 1/2 inch hole. Clamp the vise without moving it.

4. Set the correct speed and drill into the workpiece so that there is a slight chamfer. Do not spot drill for the 1/16 in. hole, but spot drill for all other holes. You might enlarge the 1/16 in. hole oversize.

5. Change to the 1/16 in. drill and drill that hole.

6. Change to the 1/8 in. drill and drill the 1/8 in. hole. Pilot drill all the others.

7. Continue drilling all the remaining holes, taking care to change to the correct speeds for each drill size.

8. Using the countersink, lightly chamfer each hole on both sides of the gage.

9. The reference marks for measuring drill lip lengths may now be put on with a scriber and a rule, a specially prepared stamp, or a scriber and a division setup on a milling machine. Check with you instructor for the method you should use.

10. Turn in the grading sheet and the finished project for your instructor's evaluation.

PROJECT 2. TURNING and THREADING PROJECT

INSTRUCTOR EVALUATION

NAME_____DATE_____

Grade

	Letter	Percent
1. Follows drawing (dimensions and tolerances)	_____	_____
2. Machining finishes	_____	_____
3. Mechanism or tool operates satisfactorily	_____	_____
4. General workmanship	_____	_____
Total grade	_____	_____

Comments

Instructor signature: _____

PROJECT 2. TURNING AND THREADING

Objectives

1. Learn to use precision measuring tools.
2. Learn to grind lathe tools for turning and threading.
3. Be able to use a small engine lathe for external turning, grooving, and threading.

Outline for Study

Prior to starting each procedure for this project, study and complete applicable units and Post-Tests for:

1. Precision measurement: **Section C**
2. Introduction to the lathe. **Section I**
3. Grinding a lathe tool: **Section I**
4. Facing and center drilling: **Section I**
5. Turning between centers, undercutting, and chuck work: **Section** I
6. Cutting threads: **Section I**
7. Measuring threads: **Section I**

Procedures

Begin the procedures for this project by completing the following worksheets.

Worksheet 1. Grinding a lathe tool bit.
Worksheet 2. Center drilling and facing.
Worksheet 3. Measuring threads.

WORKSHEET 1. Grinding a Lathe Tool Bit

Grind one acceptable right-hand turning tool bit as shown in Section I

WORKSHEET 2. Center drilling and Facing

Tools and Material

Number 3 center drill. Piece of 1-1/4 in. diameter mild steel HR round 7 1/8 in. long. Right-hand turning tool

Procedure for center drilling

1. Set up the 1-1/4 in. diameter workpiece in a chuck on a lathe and face one end without leaving a center stub.
2. Set up a Number 3 center drill in the drill chuck on the tailstock and set the correct speed. Using cutting oil, drill the center hole to the desired depth. About three-quarters of the way into the countersink section would be a good depth in this case.
3. Break the corner with a file; that is, deburr the sharp outer edge left by machining.
4. Turn the part around in the chuck and repeat steps 1, 2, and 3.

Procedure for Turning between Centers

The specifications and tolerances as given in the drawing below must be observed and held unless your instructor has a different set of allowances.

1. Set up the lathe for turning between centers. Make sure the centers and sockets are clean and the centers are well seated.
2. Apply high pressure lubricant to the center hole at the tailstock end. Ordinary oil will not work as it will run out and the center will heat up and be damaged.
3. Slip on the dog with the bent tail pointed toward the drive plate.
4. Place the workpiece between the centers.
5. Put the dog in place and tighten the set screw. Make sure the bent tail is clear of the bottom of the slot

6. Adjust the dead center so there is no end play, yet with sufficient clearance so the bent tail of the dog is free to move when the part is rotated. Tighten the tailstock binding lever. <u>Note:</u> Further adjustment of the dead center will be necessary when the work heats up from machining.

7. Fasten a right-hand turning tool in a straight or left-hand toolholder or in any appropriate holder provided on your lathe. Adjust the holder so the tool will not gouge the work if it should slip from heavy cutting. Set the tool for height.

8. Set the lathe to the correct RPM for the work.

9. Adjust the feed for roughing; 0.008 to 0.012 in. would be appropriate for a small lathe.

10. <u>Always check to see that the workpiece or dog is clear by turning the spindle by hand before using the power. Wear safety glasses.</u> Turn on the machine.

11. The first roughing cuts should extend past the center section of the workpiece so that a distance of roughly 4 1/2 in. should be turned from the tailstock end.

12. Touch the tool to the work, zero the crossfeed dial, and take a skim or clean-up cut of about 0.005 in. depth for about 1/2 in. Turn off the machine. Leave the dial setting as it is.

13. Check this turned size with your micrometer.

14. In most turning operations, it is standard practice to leave all diameters of a turned part slightly oversize to allow for any warping before taking the final finish cut. This operation is called "roughing." About 0.030 in. is sufficient to leave for finish. Most small lathes should have sufficient power to make the first diameter with one roughing cut and one or two finishing cuts. For example, let us assume the 1-1/4 in. diameter is now 1.245 in. and the finish diameter will be 1.125 in. Thus, 0.120 in. is the total amount of stock to be removed from the diameter of the middle section. About 0.030 in. should be left for finishing, leaving 0.090 in. for the roughing cut.

15. In this case, you would now set your depth of cut for 0.045 or 0.090 in. diameter, depending on the type of graduation on the cross feed collar of your lathe. Complete the roughing cut for a length of about 4k in. Check for taper and, if it is more than 0.001 in., reset the tailstock. (See Section I, "Alignment of. Lathe Centers.")

16. The next size is a 1 in. diameter that has a shoulder. All diameters are roughed on both ends before taking any finish cuts.

17. Scribe or mark a line on the workpiece 3 in. from the tailstock end.

18. Make a roughing cut to within 1/16 in. of the line. This allowance is left for later clean up of the shoulder. Increase lathe speed to compensate for smaller diameter.

19. Scribe a line 2 in. from the tailstock end and repeat step 18. Scribe a line 1 in. from the tailstock and repeat step 18.

20. Turn the workpiece end for end and rough turn each step as before.

21. Set up a right-hand finishing tool or a modified right-hand turning tool (by grinding a larger nose radius on it) in the toolholder.

22. Set the lathe for a finishing feed and speed.

23. Run the tool in until it touches the workpiece and move it to the starting position. Add 0.003 in. more and zero the micrometer collar. Take a light cut for about 1/4 in.

24. Check the diameter with the micrometer caliper. It should now be between 0.020 and 0.030 in. oversize. set the cross feed micrometer collar for the appropriate amount and take a trial cut 1/8 in. long. Check for size with a micrometer and adjust if necessary, so that the finish size will be within the tolerances given on the drawing. Finish the cut.

25. When all of the diameters are finished on one end, turn the tool so that a square shoulder can be made. Pick up the cut (just touch the finish diameter) near the shoulder and hand feed to the line. Back out the tool with the cross feed handle to square the shoulder. Turn the tool SCEA about 45 degrees and chamfer the edges.

26. Turn the work end for end and repeat.

Procedure for Undercutting

Before the threading operation is begun, a means of clearing the threading tool should be arranged. Undercutting to the depth of the thread is one such means.

1. Using either a parting tool or a specially ground tool, make an undercut for each thread equal to its single depth plus 0.005 in. Use cutting oil. Make the undercuts 3/16 in. wide. If the tool chatters, check for too much toolholder or tool overhang; slow down the machine.

2. The following formula will give you the single depth for undercutting Unified threads:

$$d = P \times 0.613 \quad \text{or} \quad d = .613/n$$

where

d = single depth
P = pitch
n = number of threads per inch (TPI)

Since there is no undercut for the 1-1/8-12 thread, consider the other six diameters and threads only.

Fill out the following list for the single depth of undercutting on each:

14 TPI _____   8 TPI _____

16 TPI _____   10 TPI _____

20 TPI _____   13 TPI _____

Use this list as a guide for depth of undercutting as you undercut each step on the project shown below.

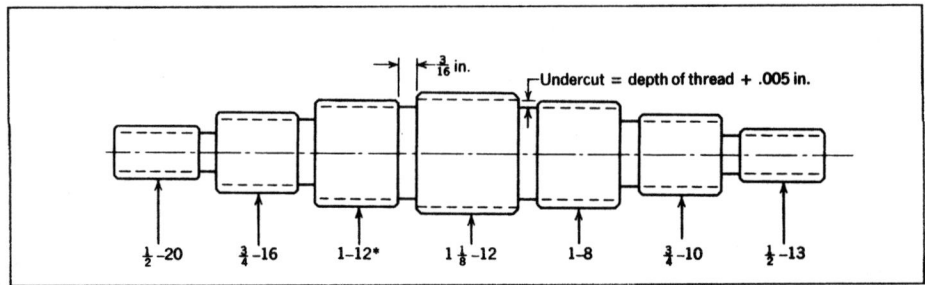

* Note: Use 1-12 UNF or 1-14 UNS (former American Standard threads). Many shops do not have 1-12 UNF nuts in stock, but have only 1-14 UNS nuts to use for testing the thread. Check with your instructor to see which thread pitch you should use.

3. Set up a parting tool and touch it to the finish diameter you wish to undercut. Zero the micrometer collar and back off the tool Now move the tool to the shoulder and make the cut. You may need to double the calculated depth on some lathes that read directly for diameters. For single depth reading cross feed collars, use the figures as they are. You may have to take two cuts side-by-side to get the 3/16 in. width.

Procedure for Threading:

1. For this operation, you will need to grind the opposite end of your lathe tool bit to make a 60 degree threading tool as shown in **Section I** Use a center gage to check it. Provide for a zero or neutral back rake on the tool, depending on the type of toolholder you are using.
2. Set the compound at 29 degrees to the right, off the cross slide index tine.
3. Determine the depth the tool is fed into the work when the compound slide is set at 29 degrees for the following threads. The formula for Unified threads is:

$$d = P \times 0.708 \quad \text{or} \quad d = 0.708/n$$

Fill out the following table for the infeed when threading each step.

12 TPI _____ 8 TPI _____

14 TPI _____ 10 TPI _____

16 TPI _____ 13 TPI _____

20 TPI _____

4. Set up the threading tool, making sure it is on center and square to the work axis. Use the center gage for aligning the tool to the work.
5. Select a slow RPM and set the quick-change gears to give you 12 threads per inch so you can do the 1 in. diameter thread first.
6. Put the feed change lever or its equivalent in neutral.
7. Set the compound micrometer collar on zero.
8. Bring the tool in contact with the work on the largest diameter using the cross slide. Then set its dial on zero.
9. Move the carriage to the right until the tool is clear of the work; feed in the compound 0.003 in. Apply cutting oil to the work. <u>Always</u> use cutting oil when threading.
10. When the index mark on the thread dial is in line with the proper position on the dial, engage the half-nut lever. Keep your hand close to it so you can disengage the lever as soon as it has cut the full length of the thread.
11. With the cross slide, back the tool away from the work one or two turns. Return the carriage by hand to the right, where you started.
12. Return the cross slide to its zero position.
13. Feed the compound another 0.005 in (single depth); 0.010 in. (double depth).
14. Repeat steps 10, 11, 12, and 13 until 0.010 in. remains of the infeed dimension. Then feed only 0.002 in. per pass. The last pass should be made without additional feed in order to clean up the thread. Check the thread for size. When you are nearing the last few passes, use a standard nut for testing the thread. It should fit snugly, but you should be able to turn it on with your fingers. When turning between centers, using a dog, the workpiece may be removed for testing without disturbing the tracking of the thread if the dog is returned to the same slot.

15. Repeat steps 5 to 14 <u>with the changes made for each specific thread.</u> When reversing the workpiece between centers, protect the thread with two nuts locked on each other, or use a threaded dog.

WORKSHEET 3. Measuring Threads

Tools and Material

Screw thread comparison micrometer

Precision plug gage

Three-wire set and micrometer

Screw thread micrometer

Finished threading project

Procedure

1. Measure all of the thread diameters and pitches.
2. Record the type of measuring system you are using and the measurement of each thread.
3. Record the correct measurement for each thread and note the difference.

Type of measuring instrument: _____

TPI	13	10	8	12	14	16	20
Your measurement							
Correct measurement							
Total error oversize							
Total error undersize							

4. Turn in the grading sheet and the finished project for your instructor's evaluation.

PROJECT 3. MAKING a CENTER PUNCH

INSTRUCTOR EVALUATION

NAME_____DATE_____

	Grade	
	Letter	Percent
1. Follows drawing (dimensions and tolerances)	_____	_____
2. Machining finishes	_____	_____
3. Mechanism or tool operates satisfactorily	_____	_____
4. General workmanship	_____	_____
Total grade	_____	_____

Comments

Instructor signature: _____

PROJECT 3. MAKING A CENTER PUNCH

Objectives

1. Learn to identify ferrous and nonferrous metals.
2. Learn to turn a taper using the compound rest.
3. Learn to make a knurl.
4. Be able to harden and temper plain carbon tool steel correctly.

Outline for Study

Prior to starting each procedure for this project, study and complete Post-Tests for applicable units in:

1. Identifying metals: **Section D**
2. Machining the center punch: **Section I**
3. Hardening and tempering the punch: **Section D**
4. Noncutting hand tools: **Section B**

Procedures

Begin the procedures for this project by completing the following worksheets in this workbook.

Worksheet 1. Identifying metals.

Worksheet 2. Machining the center punch.

Worksheet 3. Hardening the center punch.

Worksheet 4. Tempering the center punch.

WORKSHEET 1. Identification of Various Metals

Materials

Box of numbered specimens and a similar set of known and marked specimens

Procedure

Correctly identify by comparison, using various tests described in the textbook.

Item Number	Test Used	Kind of Metal
1.	_____	_____
2.	_____	_____
3.	_____	_____
4.	_____	_____
5.	_____	_____
6.	_____	_____
7.	_____	_____
8.	_____	_____
9.	_____	_____
10.	_____	_____

WORKSHEET 2. Machining the Center Punch

Material and Equipment

Engine lathe and tooling

3/8 diameter drill rod

Aluminum protective sleeve or 3/4 in. diameter rod from which to make one.

Procedure

1. Make the center punch (Figure 1) turning all the tapers using the compound rest method.

2. Make an aluminum protective sleeve (Figure 2) if one is not available. It is needed to protect the knurled surface of the of the punch from the chuck jaws when the taper is turned. The material for this is aluminum bar stock 3/4 in. diameter x 1-1/2 in. long, with a 3/8 in. diameter hole drilled in the center and then slotted with a hacksaw.

Figure 1

Figure 2

Cut off a piece of 3/8 diameter drill rod, 3/4 in. longer than the finished length of the punch. (This allows for center drilling at the ends and removing the center hole after knurling).

3. Fasten a number 2 or number 3 center drill in the tailstock drill chuck.

4. Center drill the workpiece using cutting oil.

5. Exchange the drill chuck for a tailstock center.

6. Move the work from the chuck and mark the length of the required knurl on it, then return the work to the chuck and let it extend only far enough out of the chuck so that you can knurl the marked off part.

7. Select a medium diamond knurling tool and fasten it in the tool post. Set it as close as possible to the chuck side of the compound rest with the least amount of overhang. See applicable units in Section I for knurling tips.

8. Move the carriage toward the chuck until the knurling tool just clears the chuck jaws. Turn the chuck by hand to check for clearance.

9. With the lathe turned on, set your depth of cut. (Try to get a finished knurl with the first or second pass; drill rod work hardens.)

10. Secure the knurled piece of drill rod in the protective collet (as shown in Figure 2) in the three-jaw chuck. Let 2-1/2 in. inches extend from it.

11. Select and set the correct RPM (the cutting speed for high carbon steel in the annealed condition should be about 70 SFM). Note that this RPM will be quite high, perhaps to top speed of the lathe. If you try to turn this slender rod at a low speed, it will most likely ride up on the tool and bend.

12. Fasten a right-hand tool in your tool post, so you can face off your workpiece and remove the existing center hole. <u>Caution:</u> This is not a very rigid setup; work that is supported only on one end should not normally extend from the chuck more than four or five times its diameter.

13. Face the center punch so the distance from the end to where the knurl starts is 2 in. or less. Feed by hand using the cross slide; take very light cuts.

14. The 90 degree angle is machined at this point. Swivel the compound rest to 45 degrees. Tighten the carriage lock bolt. Take light cuts (0.020 to 0.030 in. deep). Hand feed by turning the compound rest handle until a sharp point is formed. Slow, even feed makes a smooth finish.

15. The 5/16 in. diameter is now turned for a length of 2 in. in from the point to a shoulder.

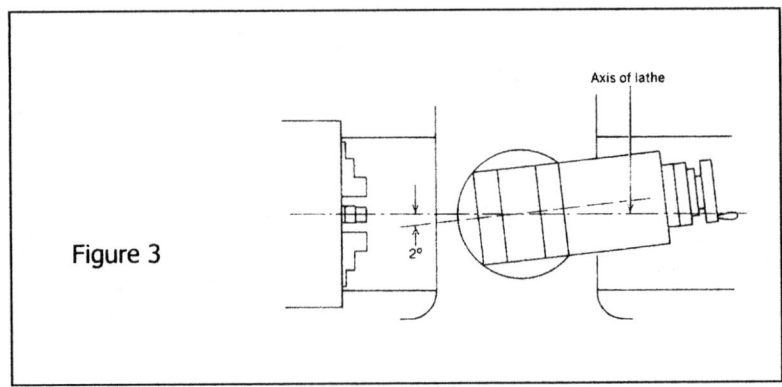

Figure 3

16. Reset the compound so it will cut an angle of 2 degrees from the centerline of the workpiece (Figure 3).

17. Turn the compound rest handle counterclockwise until it is at the end its travel.

18. Move the carriage so that the tool clear the point just turned by 1/16 to 1/8 in. and lock the carriage to the ways.

19. Take light cuts and turn a taper that measures 1-7/8 in. long from the punch point. Make a slow even feed for the last cut to produce a good surface finish.

20. Remove the workpiece from the chuck and mark the other end for the length of the 5/16 in. diameter and then knurl the length.

21. Re chuck the center punch so that the mark is visible outside the collet and turn the 5/16 in. diameter to the mark.

22, Face to the required length.

23. Chamfer the edge, his completes the machining of your center punch.

Note: The center punch must be hardened and tempered before it can be used. See the next two worksheets and follow the procedures there for heat-treating your punch. Show the completed center punch to your instructor before heat-treating it.

WORKSHEET 3.   Hardening the Center Punch

Materials

Electric or gas furnace   Quenching media (oil or water)
Tongs   Safety equipment (Face shield and gloves)

Procedure

1. Since the center punch is small tool, it should be placed in a furnace that has already been brought up to the correct temperature.

2. Determine the best place to grip the art with the tongs so that it will not be damaged when it is red hot.

3. Make sure hat you can grasp the part in the furnace to remove it in the proper orientation to enable you to quench it straight in.

4. Heat the end of the tongs so that will not remove heat from the punch.

5. When the punch has become the same color as the furnace bricks, remove it and **IMMEDIATELY** quench it **COMPLETELY UNDER** in the bath.

6. Agitate it up and down or in a figure eight motion.

7. Be sure it has cooled below 200°F (93°C) before you remove it from the quench. <u>Note.</u> If no furnace is available, a torch may be used with a color temperature chart.

Conclusion

1. Did the punch get hard? Test with an old file in an inconspicuous place. If a hardness tester is available, a quick test could be taken. (For information on hardness testing, see the unit on Rockwell and Brinell hardness testers.)
2. Is the punch as hard as it should be ? If not, check with your instructor.

WORKSHEET 4. Tempering the Center Punch

Material

| Fine emery cloth | Electric or gas furnace or oxy-acetylene torch |
| Tongs | Safety equipment |

Procedure for Tempering in the Furnace

1. Polish all smooth surfaces of the part with emery cloth and remove all oil. A cold furnace should be brought up to the correct temperature.
2. The center punch is then placed in the furnace for about 15 minutes.
3. Remove and cool in air.
4. When this method is used, the striking end of punches, chisels, and other striking tools should be given additional heating with a torch until they are a blue color on that end. This is done to ensure the safety of the user by preventing the struck end from shattering, since it is softer when tempered to blue.

Procedure for Tempering with a Torch or Hot Plate

1. When using a torch or hot plate, make sure that heat is applied to the body of a punch or part of the tool that can be softer. Allow the heat to travel slowly out to the cutting edge. In this way the colors may be observed.
2. See that the striking end of a tool is blue or gray before the proper color arrives at the cutting end.
3. When the proper color has arrived, quickly cool the piece in water. Do not delay or it will be overtempered.

Conclusion

1. Are the colors right according to Table D-7 "Temper Color Chart", in the textbook?
2. If possible, recheck hardness and compare with Table D-6.
3. Leave the temper colors on your project so your instructor can evaluate it.
4. Turn in the grading sheet with your project.

PROJECT 4. C-CLAMP

INSTRUCTOR EVALUATION

NAME_____DATE_____

Grade

	Letter	Percent
1. Follows drawing (dimensions and tolerances)	_____	_____
2. Machining finishes	_____	_____
3. Mechanism or tool operates satisfactorily	_____	_____
4. General workmanship	_____	_____
Total grade	_____	_____

Comments

Instructor signature: _____

PROJECT 4. C-CLAMP

C-CLAMP

Objectives

1. Learn to use semiprecision layout methods.
2. Learn to use a vertical bandsaw correctly.
3. Be able to drill, tap, and cross drill on a drill press.
4. Be able to do turning and threading operations.

Outline for Study

Prior to starting each procedure for this project, study and complete Post-Tests for applicable units in:

1. Layout procedures for the body: **Section E**
2. Making the body: **Section G and Section H**
3. Making the screw: **Section I**
4. Making the swivel: **Section I**

Procedures

Begin the procedures for this project by completing the following exercises and worksheets in this workbook.

Worksheet 1. Making the body.
Exercise 1. Preparing a bandsaw blade.
Exercise 2. Using the vertical bandsaw.
Worksheet 2. Making the screw.
Worksheet 3. Making the swivel.

WORKSHEET 1. Making the Body

Materials

4-1/16 IN. of 3/4 x 2 in. CR flat bar

4-1/16 in. of 1/2 in. dia. CR shaft

About 2 in. of 3/4 in. dia. CR shaft

2 in. of 3.16 dia. CR rod

EXERCISE 1. Preparing a Bandsaw Blade

1. Obtain a worn out band blades and cut several suitable lengths. Practice end grinding, welding and annealing and weld grinding procedures.

2. Practice folding band saw blades into small cols suitable for storage.

EXERCISE 2. Using the Vertical Bandsaw

1. Us the job selector and list all of the pertinent sawing data for the following materials:

   (a) Mild steel   (b) Aluminum   (c) Cast iron

2. Familiarize yourself with the band machine controls.

3. Obtain some scrap material. . Consult the job selector and set proper band velocity. Make several straight and contouring cuts to familiarize yourself with band machine operation.

4. Use the band machine to cut out your C-clamp after is it laid out.

Procedure for Making the Body

1. Obtain 3/4 x 2 in. material and cut off a 4-1/16 in. length in the power saw. Deburr.

2. Clean the material in solvent.

3. Using a vise with jaw caps, file both ends square and flat to 4 in. long. Note that all dimensions are given in Drawing II at the end of this project.

4. Apply Layout to one side. Let it dry.

5. Follow the layout procedure as shown in Section E in your text book.

6. Layout the lines where you will make saw cuts. Deepen the layout punch marks with a center punch.

7. Drill the 1/2 in. holes.

8. Using the vertical band saw, cut along the layout lines, leaving about 1/32 in. of material for filing.

9. File to the marks , keeping the work square and flat. Carefully finish file to the point of tangency of the 1/2 holes. Be careful not to nick the radius with the edge of the file.

10. File the large angles on the corners flat and square.

11. File a 1/16 in. chamfer on all the edges except for the anvil which is deburred only.

12. Lay out the 3/6-24 UNF threaded hole.

13. Follow procedures given in applicable units in **Section H** in your text book for drilling and tapping the 3/8-24 UNF hole.

Note: An alternate method to that described in the text book of holding the tap in the drill press is to clamp the tap directly in the drill chuck and start it by hand turning the chuck. **Do not turn on the power**. When you have completed several turns of the tap, release the tap shank from the drill chuck, leaving the tool still in the workpiece, take the work to a bench vise or leave it in the drill press vise and complete the tapping by hand. This will insure a straight tapped hole. Clean up the body and the threads.

WORKSHEET 2. Making the Screw

Material (CR = cold rolled)

CR shaft 1/2 in. dia. 4-1/8 in. long   2 in. of 3/16 in. CR rod

Procedure for Making the Screw

1. Take the 1/2 in. CR round; lay out the hole location and punch. Place it in one or two vee-blocks, depending on their size. Lightly clamp with about 1 in. extended from one end.

2. Set up the punch mark so it is centered and on top by using the combination square and a rule. Refer to applicable units in **Section H**

3. With a wiggler, locate the punch mark directly under the spindle.

4. Center drill; change to a 3/16 in. drill and drill through. Chamfer both sides lightly. This part is now ready for the lathe work.

5. Chuck the material in a three-jaw chuck and center drill the end (the end that is not cross drilled) with a center drill. (Since this small drill is very delicate, use a high speed and a very light feed.) A number 3 center drill will work if you do not go too deeply.

6. Reverse the work in the chuck and machine the radius on the cross drilled end. Check with a radius gage; file to finish the shape of the radius.

7. Again, reverse the work and grip the radius end with the three-jaw chuck for about 1/4 inch of the stock. Place the dead center in the other end. Use center lube.

8. Machine to size and make the shoulders to length. Turn an undercut for terminating the thread
9. Cut threads 3/8-24 UNF to proper depth with a single point tool, not with a die. Check with the tapped 3/8-24 UNF thread in the C-clamp body.

Procedure for Making the Pin

1. Saw off 2 in. of 3/16 in. CR rod. Make a full radius on each end.

2. Knurl or prick punch at the center to make a tight fit in the 3/16 in. hole. Insert the rod and tap in place.

WORKSHEET 3. Making the Swivel

Materials

Approximately 2 in. long piece of 3/4 in. diameter round stock

Procedure for Making the Swivel

1. Set up in a three-jaw chuck.

2. Machine the swivel end as shown on Figure 1 and part it off. Deburr.

Figure 1. Clamp thread swivel

Procedure for Assembling the C-Clamp

1. Assemble the clamp screw in the clamp body.

2. Place swivel end on the 1/4 in. diameter of the screw

3. Place a 5/16 in. bearing ball in the hole formed by the center drill. (This center drilled hole must be large enough to leave a very thin edge so it will roll over from the pressure the bearing ball.) Place a piece of scrap metal on the C-clamp anvil to protect it, and apply pressure by turning the screw. This procedure enlarges the end of the screw, thus retaining the swivel, yet allowing it to turn.

4. Turn in the grading sheet and the finished project for your instructor's evaluation.

C-CLAMP DRAWING 1

C-CLAMP DRAWING II

PROJECT 5. PAPER PUNCH

INSTRUCTOR EVALUATION

NAME_____DATE_____

	Grade	
	Letter	Percent
1. Follows drawing (dimensions and tolerances)	_____	_____
2. Machining finishes	_____	_____
3. Mechanism or tool operates satisfactorily	_____	_____
4. General workmanship	_____	_____
Total grade	_____	_____

Comments

Instructor signature: _____

PROJECT 5 PAPER PUNCH

Objectives

1. Learn to drill and ream holes in alignment.
2. Be able to tap threads in blind holes.
3. Learn countersinking and reaming techniques.
4. Be able to turn a large radius, using a lathe.
5. Be able to cut a thread with a hand die.
6. Be able to wind a spring by hand.

Outline for Study

Prior to starting each procedure for this project, study and complete Post-Tests for applicable units in:

1. Making the body: **Section H, Section B**
2. Making the base: **Section H**
3. Making the punch: **Section B**
4. Hardening the base and punch: **Section D**

Procedures

Begin the procedures for this project by completing the following worksheets.

Worksheet 1. Making the body and base

Worksheet 2. Countersinking the base
   Exercise 1. Identifying machine reamers

Worksheet 3. Reaming the punch hole and 1 inch hole
   Exercise 1. Identifying hand reamers

Worksheet 4. Making the punch

Worksheet 5. Making the mushroom cap

Worksheet 6. Making the spring

Worksheet 7. Hardening and tempering the punch and base

WORKSHEET 1. Making the Body and Base

Tools and Materials

Small counter bore
No. 7 drill
Center drill
1/4 in. drill
3/16 in. drill
31/32 in. drill

2-1/16 in. of 1/2 x 1-1/2 in. CR bar stock

2-9/16 in. of 3/16 x 1-1/2 in. chrome spring wire

Procedure

1. Layout the body (Figure 1) and the base (Figure 2) and punch mark where the bolt holes will be drilled except for the punch guide hole. It will be drilled later when the body and the base are bolted together. This procedure will insure perfect alignment of the punch and die hole in the base.

Figure 1.

Figure 2.

2. Set up the body (bottom up) in a drill press vise.

3. Spot the center punch mark opposite the punch hole wit the center drill and set u the # 7 drill in the chuck. Set the depth stop so that the #7 drill; will not go past the layout line of the one inch hole.

4. Drill to depth. Countersink slightly for tapping. This hole should now be tapped with a 1/4-20 UNC tap. (Refer to applicable units on tapping in **Section B**.) Do not drill and tap the other hole until the base is clamped on.

5. Take the body out of the drill press vise and place the base material in the vise.

6. Drill both bolt holes in the base (not the punch hole) with a #7 drill

7 Now, drill out the one that corresponds to the hole that you have just tapped with a 1/4 in. drill size.

8. Using a 1/4-20 UNC x 1/2 in. cap screw for a temporary fastener, bolt the base to the body and align them.

9. Place this assembly in a drill press vise, bottom up.

10. Using the #7 drill, set the depth stop and then drill the other hole in the body using the #7 hole in the base as a guide. The procedure you have just used to insure alignment of the two holes in an assembly is commonly used by machinists because it is more accurate than separately laying out and drilling two pieces that must fit together.

11. Remove the bolt holding the base and tap the second hole in the body.

12. Drill the second hole in the base to 1/4 in. dia.

WORKSHEET 2. Counterboring the base

Tools and Equipment

5/8 dia. 82 degree countersink
Drill pres and tooling
Cutting oil

Procedure

1. Set the drill press for the lowest speed, as the base is a tool steel that will rapidly work harden at higher speeds.

2. Set up the base in a vise and chuck the countersink.

3. Apply surfurized cutting oil.

4. Turn on the machine and begin cutting the countersink. Always produce a chip. If the tool rubs, the result will be work hardening of the surface and starting a new cut will be difficult.

5. When you are approaching the correct depth, stop cutting and check the fit with a 1-1/4 in. flat head screw, by dropping it into the hole. It should be flush with the surface or slightly below (about 0.010 to 0.015 in.). The size may also be checked by using tables of screw head sizes found in handbook tables.

EXERCISE 1. Identifying Machine Reamers

1. Obtain a prepared kit of reamers or ask your instructor to direct you to a place where reamers are stored , if they are not readily available to you.

2. Practice naming the different kinds of machine reamers until you can easily identify and name any reamer when you see one.

3. Select the 1/4 in. chucking reamers for the next worksheet.

WORKSHEET 3. Reaming the Punch Hole and the One Inch Hole

Tools and Equipment

1/4 in. dia. fluted chucking machine reamer
1 in. dia. shell reamer
Appropriate drills
Sulfurized cutting oil

Procedure

1. At this point, the holes should all have been drilled with exception of the punch hole.
2. Bolt the body and base together with the flat head screws and set up in the drill press (Figure 3).
3. Spot the layout punch mark on the top of the body and drill through the body and the base together.
4. Drill through both body and base with a letter C drill.
5. Maintaining the setup, ream to 1/4 in. diameter.

Figure 3.

Figure 4.

6. Now, drill a 5/16 in. diameter hole part way into the base from the bottom to allow for paper chaff clearance.
7. Set up and align for cross drilling the 3/32 in. hole through the 1/4 in. hole. Be sure to plug the 1/4 in. hole with a soft steel round. It may be necessary to mark the plug a few times with a punch to make a tight fit in the 1/4 in. hole. Align the spindle with the layout punch mark and chuck a 3/32 in. drill. Drill through.
8. Using an upright or radial arm drill press, set up for drilling the 31/32 in. hole in the body. Spot drill, then use a large pilot drill followed by the 31/32 in. drill size.
9. Maintain setup and machine ream this drilled hole to 1 in. dia. (Figure 4).
10. Using a large 45 degree countersink, chamfer the edge of the 1 in. hole 1/32 in. deep.
11. The notch on the bottom of the body may now be sawn out and filed to the layout line.

EXERCISE 1. Identifying Hand Reamers

1. Ask your instructor to direct you to the location where a variety of reamers are stored. With the aid of this exercise, identify the different kinds of hand reamers. Practice this until you can identify and name a hand reamers when you see one.

2. Hand ream the 1/4 in. diameter punch hole with an expansion reamer so that the punch has a sliding but minimum clearance fit.

WORKSHEET 4. Making the Punch

Material

2-9/16 in. of 1/4   O-1 or W-1 drill rod

Procedures

1. Saw the 2-9/16 in. length of 1/4 in. drill rod, using a hand hacksaw and vise jaw caps.
2. Chamfer one end and thread with a hand die, 1/4-20 UNC x 1/4 in. long. Use sulfurized cutting oil and back up the die frequently to break up the chips.
3. File the radius on the other end as shown in Figure 5, using a 5/16 in. dia. round file. Deburr
4. File a flat, as shown on the drawing, approximately 33/64 in. deep.
5. Insert the punch into the guide hole and align the flat with the 3/32 in. hole.

Figure 5.

6. Cut off a 9/16 in. length of 3/32 in. rod and insert it into the body.
7. If the punch does not move freely along the flat, ten remove it and file it to a greater depth. Also make sure that the cutting end of the punch slides complete back into the body. If it does not, file the flat longer toward the cutting end.
8. The punch should also have a sliding fit (free, but not loose) in the hole. If the machine reamer did not provide an adequate sliding fit, then a hand reamer or an expansion reamer should be used to resize the hole and provide the correct fit.

WORKSHEET 5. Making the Mushroom Cap

Material

9/16 in. of 1-1/2 in. dia. HRMS round bar

Procedure

1. In a small lathe, using the thee-jaw chuck, set up the 1-1/2 in. dia. stock to turn relatively true (no ore than 1/32 in. axial run out).

3. Face one end and center drill. Drill 3/8 in. deep and tap 1/4-20 UNC, first with a plug tap followed by a bottoming tap. (Figure 6)

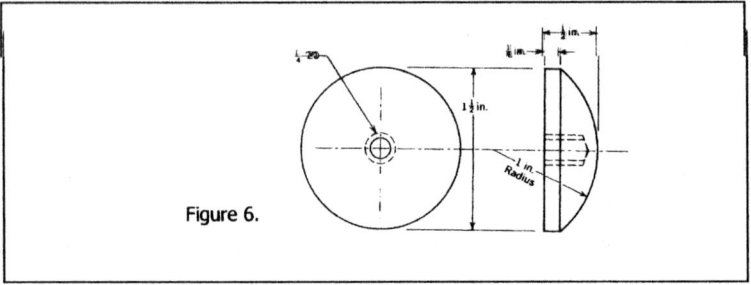

Figure 6.

3. Reverse the cap in the chuck, gripping less than 1/8 inch.

4. The 1.000 inch radius may be turned with a radius attachment or it may be done freehand. If it is done freehand, a radius gage must be frequently used to check your progress. Tool marks can be removed by filing, first with a rough or double cut file and then with a single cut file. The cap may then be finished with abrasive cloth, backing the cloth up with a file so your fingers will be away from the chuck jaws.

WORKSHEET 6. Making the Spring

Procedure

1. Use 17 gage spring wire to wind a spring about 1 in. long. Place two blocks of hardwood in the vise. Place a 1.4 in. dia. mandrel (slotted on the end) between the blocks (Figure 7).

2. Insert the end of the spring wire in the slot. Hold the wire at approximately the lead angle of the spring and turn the mandrel with the vise tight enough to place a drag on it. Wind the spring about 1 inch long.

3. When the spring is wound, pull it from the mandrel and cut or break it to length by nicking the wire on the edge of a grinding wheel. (Spring wire should NOT be cut with wire cutters). If the coils are too close together, grip the ends of the spring with pliers and pull the coils apart until the spacing is correct (Figure 8). Flatten the ends of the spring by grinding on the flat side of the wheel.

Figure 7.   Figure 8.

WORKSHEET 7. Hardening and Tempering the Punch and Base

Procedure

1. See applicable units in section D, before doing any of the heat treating

2. Heat about 3/8 in. of the cutting end of the punch with a torch to about 1500 deg. F.
(a cherry red color) and quench in oil or water, depending on the composition of the
steel, whether it be O1 or W1. The reason the punch is not hardened for its full length is the probability of warping, especially with a flat on one side.

3. Temper the punch on a hot plate.

4. The base may be hardened all over, as in a furnace and quenched in the proper medium. Be sure that all the sharp corners are chamfered (except for the punch hole).

5. Polish the base by laying it on a strip of abrasive cloth on a hard flat surface and sliding the base back and forth over the abrasive.

6. Temper the base on a hot plate.

Assembly of the Paper Punch

1. Make sure that all the parts are chamfered before assembly.

2. After you have assembled the parts your paper punch, check for correct operation. Lightly oil the punch. Check to see if it will cut through one thickness of paper cleanly.

3. If everything works well, the 3/32 in. pin may be slightly riveted to retain it in its hole.

4. The pin may now be filed flush with the body.

5. Turn in the grading sheet and the finished paper punch for your instructor's evaluation.

PROJECT 6.   PARALLEL CLAMP

INSTRUCTOR EVALUATION

NAME_____DATE_____

Grade

	Letter	Percent
1. Follows drawing (dimensions and tolerances)	_____	_____
2. Machining finishes	_____	_____
3. Mechanism or tool operates satisfactorily	_____	_____
4. General workmanship	_____	_____
Total grade	_____	_____

Comments

Instructor signature: _____

PROJECT 6. PARALLEL CLAMP

Parallel Clamp

Objectives:

1. Learn to drill two or more holes through two parts and maintain alignment for true position.

2. Learn how to make long threads on slender pars with a button die holder.

Outline for Study

Prior to starting each procedure for this project, refer to the applicable units in:

1. Making the legs and clip: **Section G**

2. Making the screws: **Section I**

Procedures

WORKSHEET 1 Making the legs and clip

Tools and Materials

21 tap drill
Letter I tap drill
Letter P drill
1/4 in. dia. drill
1/2 in. dia. drill
3/16 in. dia drill

Two pieces of 5/8 x 5/8 keystock x 4 in. long

One piece of 1/16 in. sheet steel 1 x 2 in.

Procedure for Making the Legs

1. Clamp both legs together wit the long flat sides together as they will be when assembled. Set up in the drill press vise.

2. Place the leg up that will not have the 10-32 UNF tapped hole (Figure 1).

3. Spot both layout punch marks. Chuck a 1/4 in. drill.

4. Set the depth stop so that the 1/4 in. drill will go through one leg and into the other only 1/4 in. Drill the end hole.

5. Drill the other hole through both legs.

6. Remove the setup and unclamp the legs.

7. Set up the leg with the 1/4 in. hole that is 1/4 in. deep (Figure 2). Drill the hole for the 10-32 UNF tap using a #21 drill. Drill the center 1/4 in. hole out to an F drill size.

8. Set the other leg (Figure 1) and drill it through both legs with an I drill size.

9. Center all holes. Tapping may be done now.

10. Bolt the legs to together with the 1/4 in. hole inside and set up in the bench vise for filing the radius.

11. Rough file both legs together almost to the radius line that you have previously scribed.

12. Finish file, checking the radius often with a 5/16 in. radius gage.

13. Remove your filing setup when finished and carefully lay the two legs together as they will be when the clamp is assembled. (They should already be bolted together in this position.)

14. Apply layout dye to the sides where you will scribe the angular lines.

15. Scribe the lines. Make sure that they are on the correct side of the legs.

16. Saw about 1/32 in. away from the lines with a hacksaw or on the vertical band saw. If you use the band saw, use a push stick and have the correct band and cutting speed setting for the material.

Figure 4.

6. Scribe a mark 3-1/16 in. from the end of the tailstock end and with a 1/16 in. wide tool bit, make a groove 1/16 in. deep next to the scribe mark.

7. Make the 1/8 in. radius groove with a round nose tool. Check often with a radius gage while it is being cut. Use cutting oil.

8. Turn the 5/16 in. diameter for a length of 3.00 in. Chamfer sharp corners.

9. Set up a 5/16- 24 UNF die and stock, or button die holder if one is available. Use the tailstock spindle as a guide to start the die if a die stock is used. Refer to applicable units in section H in the textbook.

10. Using slow speed, bring the die and holder to the work. Use cutting oil and start the die. When you have made about 1/2 in. of threads, stop the lathe and back up the die to break the chips. Repeat this operation until the threads are completed.

11. Set up the parting tool and begin cutting off the screw 4 in. from the end. Part off.

12. Turn the material end for end and chuck about 3/4 in. of the end. Measure the total depth of the lathe center hole and record it (so you can remove that amount later). Lubricate the center hole and adjust the dead center. You are now ready to proceed with the screw that has the swivel end (Figure 5).

Figure 5.

13. Since the center hole will be removed lather on this screw, extra length must be allowed on the threaded portion for this operation. Knurling should begin as near as possible to the chuck and to within 3 in. of the end.

14. The 1/8 in. radius groove is made as before, after you have done the layout.

15. Turn the 3/16 in. diameter and the 1/4 in. diameter on the end.

16. Cut the 5/16-24 UNF thread with a die.

17. With the two legs clamped together in a bench vise, rough file the angles and then finish file. Your instructor may want you to mill these surfaces.

18. File a 45 degree by 1/16 in. chamfer on the angles and outer edges and around the radius. Do not chamfer the clamping side except for lightly deburring it to remove the sharp edges.

19. Case hardening is not recommended as the legs tend to warp when quenched.

20. If gun bluing or machinery black compounds are available in your shop, you may wish to put a finish on the legs.

Procedure for Making the Clip

1. The material used should be larger than the finished size because of the difficulty holding and drilling the part. The 1/2 in. hole should be drilled first for this reason.

2. The off set should be made in a bench vise by placing two 1/16 in. shims on opposite sides of the lip in an offset position. When the vise is tightened, the clip will be bent offset.

3. Layout and drill the 3/16 in. hole. The clip may then be finished (Figure 3).

Figure 3.

WORKSHEET 2. Making the Screws

Tools and materials

Lathe
Hand die
Knurling tool

Parting tool
Grooving cutter

10 in. of CF round bar stock

Procedure

1. The 10 in. of 5/8 in. dia CF round stock will give you sufficient materials to make the two screws for the parallel clamp with extra length` to hold in the lathe chuck jaws. Use the three-jaw chuck.

2. Center drill both ends. hen one screw has been finished and parted off, the remaining piece can be turned end for end. This saves set up time.

3. Extend about 5 in. of stock out from the chuck. Lubricate and adjust the dead center. Begin with the screw wit the clip grooves (Figure 4).

4. Layout with a scribe and a rule on the circumference of the work, the positions for grooving and knurling.

5. Set up for knurling. Use a medium knurl, starting 3-1/4 in. from the end tailstock end.
Knurl about 7/8 in.

17. Set up a parting tool and begin to cut off 7/8 in. from the shoulder of the threaded section. Chamfer as before and complete the cutoff operation.

16. Screw two 5/16-24 UNF nuts about an inch onto the thread and jam them tightly together.

19. Mount the screw in a three-jaw chuck on the double nuts with the center hole extending out.

20. Set the compound to 59 degrees with the lathe centerline and turn the swivel end to 1/4 in. in length. Remove the work from the chuck and unscrew the nuts.

21. The 3/16 in. hole in the knurled knob should be drilled. The knurl must be protected from damage with aluminum of soft copper when it is clamped in a vee-block or vise.

22. The parallel clamp may now be assembled.

Assembly of the Parallel Clamp (Drawing 1)

1. The parallel clamp is assembled by turning the correct screws in their legs and by fastening the clip with a 10-32 UNF fillister head screw x 3/8 long.

2. Run the legs of your parallel clamp together. hey should come together easily and squarely.

3. Turn in the grading sheet and completed parallel clamp for your instructor's evaluation.

NO.	DESCRIPTION
1	SCREW
2	CLIP
3	FLL. HD. SCREW 10-32
4	LEG
5	LEG
6	SWIVEL SCREW

PARALLEL CLAMP
MACHINE TOOL PRACTICES

PARALLEL CLAMP DRAWING 1

PROJECT 7. TAP HANDLE

INSTRUCTOR EVALUATION

NAME_____DATE_____

Grade

	Letter	Percent
1. Follows drawing (dimensions and tolerances)	_____	_____
2. Machining finishes	_____	_____
3. Mechanism or tool operates satisfactorily	_____	_____
4. General workmanship	_____	_____
Total grade	_____	_____

Comments

Instructor signature: _____

PROJECT 7. TAP HANDLE

Objectives

1. Learn to set up and turn rectangular bar stock
2. Learn to tap threads on the lathe
3. Lean to knurl slender bar stock.

Outline for Study

Prior to starting this project, study and complete applicable units and Post Tests for:

1. Making part # 1: **Section C**
2. Making part # 2: **Section I**

Procedures

Begin the procedure for this project by completing the following worksheets:

Worksheet 1: Making Part #1
Worksheet 2: Making Part # 2

Refer to the drawing at the end of this worksheet. Tap Handle: Drawing I

WORKSHEET 1. Making Part # 1

Material

5-1/2 in. of HRMS flat bar 1/2 x 1 in.

Procedure

1. Cut off a piece of hot rolled mild steel (HRMS) flat bar 1/2 x 5-1/2 in. length

2. Clean the lathe spindle nose and mount a four-jaw chuck.

3. Fasten your workpiece in the chuck and let approximately 1 in. extend out.

4. Center your work to within .003 in., using methods shown in applicable nuts in Section I.

5. Use a right hand turning tool and face the end with a good finish. Do not remove more material than is necessary and do not leave a stub in the center.

6. Mount a # 3 center drill in the tailstock drill chuck.

7. Select and set the spindle speed for center drilling. If the speeds is too slow, the center drill may break.

8. Center drill the end. Use cutting oil.

9. Take the workpiece out of the chuck, reverse it, and repeat steps 5 through 5.

10. Remove the workpiece from the chuck.

11. Remove the chuck and substitute the drive pate and center in the headstock.

12. Fasten the dog on your workpiece and set it up for turning between centers.

13. Select the spindle speed for rough turning.

14. Select the feed rate for the roughing cut.

15. Mark your workpiece 2 in. from the tailstock end and take roughing cuts to this mark until the diameter is approximately .800 in.

16. Now, mark your workpiece 7/16 in. from the tailstock end and turn that portion to approximately .560 in. diameter.

17. Reverse your workpiece end for end, clamping the dog on the end you have just turned.

18. Mark your workpiece 1-9/16 in. from the headstock end and rough turn to that mark until the diameter is approximately .560 in.

19. Now, make a mark 2-1/16 in. from the headstock end and turn your piece to approximately .430 in. dia. to that mark. Part #1 is now all machined to oversize dimensions and is ready for the finishing operations.

20. Resharpen your tool so it can be used for finishing, or make a finishing tool.

21. Select a spindle speed for finishing. This will be slightly higher than for roughing.

22. Select a feed rate for finishing. This will be much slower than that used for roughing.

23. Mark the work at 2 in. from the headstock end.

24. Turn the 3/8 in. diameter with a tolerance of .377 to .373 in. Machine to the mark.

25. Finish turn the 1/2 in. diameter with a tolerance of .508 to .494 in. It should end exactly 1-1/2 in. from the headstock end.

26. Set up a medium diamond knurling tool in the tool post.

27. Use a relatively low speed and a feed of about .010 to .020 in.

28. Make a mark 2-13/16 from the headstock end: this is how far your knurl will go.

29. Bring the knurls to the work near the dead center and feed in until you can see a definite diamond pattern develop. If there is one solid line in one direction and short broken lines in the other, either raise the tool up, lower it, or turn it slightly sideways so it will "bite" in better. If that does not work, find a different knurling tool and try that. Use lube or cutting oil on the knurls and on the work. When a full diamond pattern develops, stop knurling, since further work will ruin the job. Two passes should be sufficient, however, this workpiece will spring in the middle so more passes may be necessary at the center of the workpiece. Remove the knurls while the work rotates to avoid bending the part.

30. Reverse your workpiece and protect the knurled section with some soft material such as aluminum when you put the dog on.

32. Change speed and feed for a finishing cut and set up your right-hand turning tool again.

33. Turn the 1/2 in. diameter nearest the tailstock end with a tolerance of .500 to .494 in. and 1/2 in. long.

34. Set the compound rest so it will cut the 30 degree angle. Also set up the tool for the cut (Figure 1).

35. Lock the carriage in place and cut the taper. Make sure you get a smooth transition between the 1/2 in. diameter and the taper.

36. Swivel the compound rest to cut the other 30 degree bevel (Figure 2)

37. Turn this taper, being careful also to make a smooth transition between the taper and the 1/2 in. diameter.

38. Use the cutting edge of the tool set at 45 degrees to the work and make the two required chamfers.

39. Remove the dog from the workpiece and install a four-jaw chuck on the spindle.

40. Mark your workpiece 5 in. long. Use a protective aluminum sleeve and chuck the workpiece with the small end out. Allow approximately 1 in. to extend from the chuck.

41. Adjust the chuck jaws until the workpiece runs true.

42. Face off the excess material (to your mark).

43. Rough turn the 3/16 in. radius and use a file to complete the shape and to finish it. Check the radius with a gage.

44. Remove your workpiece from the chuck.

45. Lay out for the 3.8 drilled hole on the flat side of the large diameter. Center punch.

46. Clamp the large diameter in a drill press vise, using soft protective material on the jaws. Use a level on the part to make sure that it is set up square to the spindle after the vise has been tightened. Check to see if the drill press table is level, if not, use the same bubble position for setting up the part.

47. Spot drill with a small center drill.

48. Drill the 3/8 in. hole and chamfer lightly on both sides with a countersink.

49. Chuck the workpiece on the 3/8 in. diameter using the protective sleeve. This time let the larger part extend out from the chuck.

50. Adjust the jaws until the work runs concentric.

51. Determine the tap drill size for a 5/16-24 UNF tap and procure this drill.

52. Hold the drill in the tailstock drill chuck and drill the part to a depth of 1 inch. The center drilled hole will act as a pilot to start the tap drill.

53. Remove this drill and fasten a 5/16-24 UNF tap in the drill chuck.

54. Loosen the tailstock clamp so it can slide freely on the ways.

54. Put cutting oil on the tap and turn the chuck and workpiece in it by hand. After you bring the tap in contact wit the work, it will pull itself into the work. Back up the spindle frequently to break the chips.

55. When the tap has reached full depth, turn the chuck in reverse by hand. This will push the tap out.

WORKSHEET 2. Making Part # 2

Material

4-3/4 in. of 3/8 in. dia. CR round stock

Procedure

1. Cut off a piece of 3/8 in. diameter cold rolled round stock 4-3/8 long.

2. Set up your workpiece in a three-jaw chuck leaving 1 in. extending from the chuck. Check it for runout. If it is excessive, use a four-jaw.

3. Center drill the workpiece. Then extend it 3 in. from the chuck.

4. Support this end with a center in the tailstock. Use center lube.

5. Put a mark on your workpiece 2-9/16 in. from the tailstock end. This is how far the knurl will extend.

6. Repeat steps 26 to 29 in the procedure for making part # 1.

7. Use a protective sleeve to hold the knurled piece so that the end with the center hole extends 1 in. from the chuck.

8. Use your right-hand tool and cut 3/6 in. off the end of the workpiece. This should remove the center hole.

9. Rough turn and use a file to make the 3/16 in. radius.

10. Remove your workpiece from the chuck and mark it 2-11/16 in. from the rounded end.

12. Center drill the end and support it with a center.

13. Turn the 5/16 in. diameter with a tolerance of .312 to .310 in. diameter to the mark.

14. Turn the 1/4 in. diameter at the end of end with a tolerance of .25 to .248 in. diameter so that 1 in. of the remains of the 5/16 in. diameter.

15. Make a small undercut 3/32 in. wide to the minor diameter of your thread.

16. Set up your threading tool and thread the workpiece. Make sure that your tool is sharp. (an alternative method is to use a button die holder in the tailstock. No undercut is needed in this case.)

17. Use your right-hand tool to cut the 1/4 in. diameter back so it will be 5/16 in. long.

18. Chamfer the 1/4 in. diameter to the specifications on Drawing I. This completes the turning of Part # 2.

Assembling the Tap Handle

1. Use a small square file and make the vee in Part #I. It should be symmetrical to the center line.

2. The short 1/4 in. end may be case hardened if desired in order to prevent it from splaying out with repeated use. This can be done with an acetylene torch. Only the small end is heated to a bright cheery red and then placed in Kasenite. It is reheated and quenched in water.

3. Assembly both pieces. Use oil as a rust preventative.

4. Check to see if it will hold a tap square to the handle.

5. Turn in the grading sheet and the finished tap handle for your instructor's evaluation.

FIGURE 1

FIGURE 2

PROJECT 8. DIE STOCK

INSTRUCTOR EVALUATION

NAME_____DATE_____

Grade

	Letter	Percent
1. Follows drawing (dimensions and tolerances)	_____	_____
2. Machining finishes	_____	_____
3. Mechanism or tool operates satisfactorily	_____	_____
4. General workmanship	_____	_____
Total grade	_____	_____

Comments

Instructor signature: _____

PROJECT 8. DIE STOCK

Objectives

1. Learn to set up rectangular stock in a four-jaw chuck for boring operations.
2. Bore and counterbore precision internal diameters.
3. Learn to use the taper attachment and perform taper calculations.

Outline for Study

Prior to starting each procedure for this project, study and complete Post Tests for applicable units in:

1. Making the body: **Section I**
2. Making the handles: **Section I**
3. Alignment of lathe centers: **Section I**

Procedures

Begin the procedures for this project be completing the following worksheets.

Worksheet 1. Testing Alignment of Lathe Centers with a Test Bar.

Worksheet 2. Testing for Taper by Turning

Worksheet 3. Making the Body
       Exercise 1. Using Measuring Tools

Worksheet 4. Making the Handles
       Exercise 1. Making a taper

WORKSHEET 1. Testing Alignment of Lathe Centers with a Test Bar

Materials

Lathe   Test bar   Dial indicator

Procedures

Set up and test for taper between centers in a lathe and adjust the tailstock.

1. Clean the headstock and tailstock spindle tapers and insert the centers

2. Place the test bar between centers and adjust the lathe centers so that there is no end play.

3. Set up a dial indicator in the tool post and adjust the contact point to the height of the lathe spindle centerline.

4. Move the dial indicator along the test bar toward the headstock. Set the indicator bezel to zero. Now, move the indicator along the test bar toward the tailstock and observe the reading on the dial. If it is not off more than 0.0005 to 0.001 leave it as it is.

5. If the reading is positive (clockwise), move the tailstock away from the operator side until the needle points to the zero index.

6. If the reading is negative (counterclockwise), move the tailstock toward the operator until the needle points to the zero index.

7. Repeat the test for a double check of your results.

WORKSHEET 2. Testing for Taper by Turning

Tools and Equipment

Lathe   Micrometer   Test piece
Dial indicator   Right hand turning tool

Procedure

Set up and test for taper between centers in a lathe and adjust the tailstock if needed.

1. Clean the headstock and tailstock spindle tapers and insert the centers.

2. Place the test piece between centers with a drive dog. Use high pressure lubricant on the center hole engaged on the tailstock center. Adjust for turning.

3. Set up your right hand turning tool. First, take a roughing cut if the test piece is irregular (not completely round). Then, take a light cut (about 0.015 in.) with a finishing feed as far as you can safely go along the full length of the test peice.

4. Measure the diameter of the test bar at both ends and subtract the smaller dimension from the larger dimension. If the difference is not more than 0.0005 to 0.001 inch per foot of test bar length, the tailstock need not be adjusted.

5. Set up a dial indicator to contact the centerline near the end of the test bar at the tailstock end. Set the indicator bezel to zero.

6. If the tailstock end is smaller in diameter, then move the tailstock away from the operator's side by half the difference in the diameters.

7. Take another light cut over the full length of the test bar. Measure both ends to double check for alignment of the lathe centers.

WORKSHEET 3. Making the Body

Materials

4-3/16 in. of 3/4 x 2 CR flat bar   Two 1/4-20 UNC socket head set screws x 1/4 in. long

EXERCISE 1.

1. Inspect your shop and tool crib and identify as many comparison measuring tools as possible.

2. Obtain the small hole gage and the telescoping gage measuring kit from your instructor and note how they work for transferring measurements.

3. If you have a cylindrical or micrometer square available, use it to check some of the combination squares in your shop. The results may surprise you.

Procedure for Making the Body (DRAWING I)

1. Place a bar of **3/4** x 2 in. cold rolled mild steel in the power saw and cut off 4-1/8 in. plus 1/16 in. for finishing, or 4-3/16 in. total.

2. Deburr the sharp edges and clean the piece in solvent.

3. Find the center of the large flat side and center punch.

4. Set up in a four-jaw chuck with the center punched side out. Two of the jaws may have to be reversed. The punch mark should be adjusted to run true with the machine axis. A wiggler and a dial indicator may be used for this or a sharp dead center in the tailstock could be used to locate the punch mark. True up the face using a dial indicator.

5. Drill through the material, using first a center drill followed by a pilot drill and a 3/4 in. diameter drill.

6. Set up a boring toolholder with a 1/2 in. bar. Set the bar to hold the toolbit in an angular position. Grind the tool to cut properly at that angle so that an internal square shoulder can be made. The bar should extend no further from its holder than necessary.

7. Using the correct RPM and feed, bore to 1-3/16 in. dia. with a tolerance plus or minus .005 in

8. Counterbore to a rough size of about .025 in. under finish diameter. Leave 1/32 approximately in the bottom of the counterbore for finish; that is, make the length of the counterbore 15/32 in.

9. Check your toolbit for builtup edge or dullness and hone if necessary.

10. Set the feed and RPM for finishing.

11. One method of turning a counterbore to depth is to touch the toolbit to the outside surface of the workpiece near the bore. Using a steel rule laid flat on the ways and touching the carriage, mark at an inch line or at the end of the rule on the lathe way with a grease or graphite pencil. (Note: These markers will not harm a precision way, but do not use a scriber or any hard or sharp marking device.) This will provide an index position for moving the carriage. Position the boring bar with the cross slide so the bar and tool can enter the bore. Another method is to use the compound micrometer collar to set the depth of the tool with the compound parallel to the center axis of the lathe.

12. Bring the tool to the inner edge of the bottom of the counterbore and set it to 1/2 in. depth. Lock the carriage.

13. Start the lathe and slowly hand feed the cross slide to finish the face to depth. Watch the tool in the bore, and when the cut is at the diameter of the large bore, stop feeding and turn off the lathe.

14. Unlock the carriage and carefully clear the tool and back it out from the work.

15. Take a light trial cut on the 1-1/2 in. dia. bore (about 0.005 in.) for a distance of 1/8 in.

16. Measure the trial cut and subtract the measurement from 1.505 in. The difference will be the depth of the finish cut as measured on the diameter.

17. If the lathe you are using is graduated on the micrometer collar to read on the diameter of the work, then simply set the number you have found in step 16. If, on the other hand, the dial is graduated for single depth, then set the depth of cut at half the value found in step 16.

18. Take another trial cut, this time for about 1/4 in. Measure and correct any error caused by tool spring. The error will generally be undersize.

19. Finish bore to full depth. Stop the power feed before the tool comes to the bottom. and hand feed to make a sharp corner.

20. Chamfer all sharp edges with a tool set at 45 degrees. Remove the workpiece from the lathe.

21. Layout for center on both ends of the material. Center punch.

22. Center drill in a drill press on both ends, or use an alternate method: Center drill in the lathe and face to length.

23. Tap drill both holes for set screws.

24. Set up for turning between centers. Alternately, you could use a four-jaw chuck and dead center.

25. Rough out both ends to 21/32 in. dia. by 31/32 in. long

26. Turn the center section to 1-7/8 in. diameter using a finishing tool and feed for the last cut.

27. Set the compound for 45 degrees and rough turn both angular surfaces to within 1/32 in. of finish size.

28. Face both ends so that the total length is 4-1/8 inches.

29. Finish turn the 5/8 in. dia. ends to finish diameter and to the 1 inch length.

30. Finish turn the angular surface with the compound set at 45 degrees. Turn the compound hand crank slowly and evenly to provide a good finish.

31. Remove the part from the lathe and install a three-jaw chuck.

32. Place the 5/8 in. end in the chuck and tighten.

33. Tap drill 3/4 in. deep for a 7/16- 20 UNF tap. Chamfer the hole.

34. Place tap in the Jacobs chuck and start by turning the three-jaw chuck by hand. Leave the tailstock loose on the ways. Use cutting oil on the tap. Continue to alternately turn the chuck forward and reverse to break the chips until the tap offers more than usual resistance, showing that it has reached the bottom of the hole. Back out the tap and repeat the process on the other end.

35. Remove the part from the lathe and tap the two holes for the set screws. Clean up the part and deburr where necessary. The body is now completed.

WORKSHEET 4. Making the Handles

Materials   10 in. of 5/8 in. dia. CR shaft

Procedure

1. Place some 5/8 in. CR round stock in the saw. Cut off two pieces 5-1/2 in. long.

2. Set up material in a three-jaw chuck. Face off and center drill both ends.

3. With one end in the chuck and the other supported with a dead center, knurl from the tailstock end to a mark 3-3/16 in. from the headstock end

4. Machine the step and shoulder for the 7/16-20 UNF thread.

5. Thread the end by single point tool method. Do not use a die. Check with a 7/16-20 UNF nut or with the body of the die stock.

6. Machine the taper using the taper attachment. You will find the formula for the various methods of producing a taper in your textbook in Section I. Determine the taper per foot and set that amount on the taper attachment.

7. Chuck the knurled end extending the end opposite the thread, using soft aluminum or copper to protect the knurl. Face the part to length.

8. Freehand cut a 5/16 in. radius. Check frequently with a radius gage. Finish with a single cut file.

9. Repeat steps 1 through 8 to make the other handle.

10. Clean all parts in solvent and wipe dry. The die stock is now ready for assembly.

EXERCISE 1. Making a Taper

Note: This exercise may be used to make the tapers on the handles if a taper attachment is not available and if you have your instructor's permission.

Tools and Material

Lathe and tooling

Procedure

(a) Make a taper plug by the offset tailstock method (Figure 1).

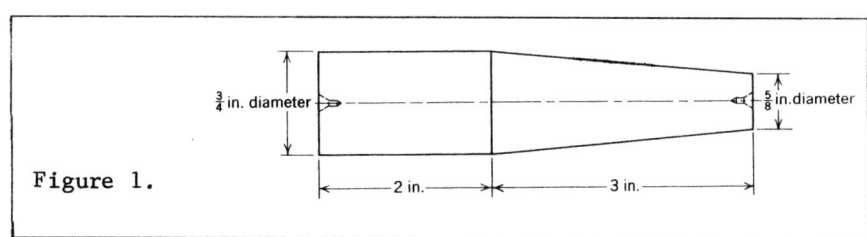

Figure 1.

(b) Measure the taper per inch.

1. Cut off a piece of CF mild steel 3/4 by 5-1/8 in. long
2. Center drill and face both ends to an overall length of 5 in.
3. Fasten a lathe dog to your workpiece and set up for turning between centers.
4. Calculate the tailstock set over and record here:
5. Use a dial indicator or the compound rest slide to offset the tailstock.
6. Take some trial cuts until the tapered portion is approximately 2 in. long.
7. Calculate the taper per inch. Record here:
8. Make two marks on the tapered part exactly 1 in. apart with a scriber.
9. Measure the diameters on these marks and subtract the smaller from the larger. This difference should be the same as the answer in step 7.
10. If it is not the same or within 0.001 in., adjust the tailstock and make trial cuts until the result is correct.
11. Mark the workpiece 3 in. from the tailstock end and make as many passes as necessary to bring the taper to that length.

Assemble the Die Stock

1. Turn in the grading sheet and finished die stock for your Instructor's evaluation.

PROJECT 9. CENTER and TAPER SLEEVE

INSTRUCTOR EVALUATION

NAME_____DATE_____

Grade

	Letter	Percent
1. Follows drawing (dimensions and tolerances)	_____	_____
2. Machining finishes	_____	_____
3. Mechanism or tool operates satisfactorily	_____	_____
4. General workmanship	_____	_____
Total grade	_____	_____

Comments

Instructor signature: _____

PROJECT 9. CENTER and TAPER SLEEVE

CENTER and SLEEVE

Objectives

1. Learn how to produce a precision taper.

2. Learn how to make a precision tapered bore.

3. Be able to fit up an external taper to an acceptable degree of accuracy.

Outline for Study

Prior to starting each procedure for this project, refer to the following applicable units in:

1. Making a Morse taper: Section I

Procedures

Begin the procedures for this project by completing the following worksheet

Worksheet 1. Making a Morse taper

    Exercise 1. Turning a Morse Taper live Center

    Exercise 2. Turning an Internal Morse taper Sleeve

WORKSHEET 1. Making a Morse taper

Exercise 1. Turning a Morse Taper Live Center

    Make a Morse taper live center (Figure 1). Your instructor may wish to assign you a certain size of Morse taper.

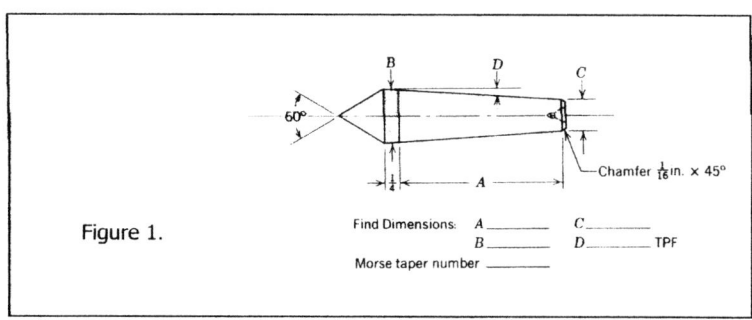

Figure 1.

Find Dimensions: A _____ C _____
B _____ D _____ TPF
Morse taper number _____

Materials

One piece of HRMS round stock 1 in. dia. by 4-1/2 long for a No. 3 Morse taper

Other taper sizes will have different dimensions that can be determined by the tables in **Section I**.

Procedure

1. Cut off mild steel round stock slightly larger than the B dimension. and 1/4 in. longer than the finished dimension of the center. The length of the 60 degree angle can be determined by multiplying .866 by the large diameter of the center.

2. Set up the stock in a three-jaw chuck and center drill the end.

3. Extend the work outward, holding only 1/2 in. in the chuck jaws. Support the other end with the tailstock center.

4. Turn the workpiece to diameter B for more than length A plus 1/4 inch.

5. Determine the taper per foot D, and record here: _____

6. Clean and lubricate the taper attachment.

7. Adjust the taper attachment for the correct TPF.

8. Move the carriage so the tool is where the taper is to be cut on the workpiece and lock the anchor bracket to the lathe bed.

9. Disengage the cross feed nut by removing the screws on the cross slide. Cover the holes to keep the chips out. This procedure is not necessary if your lathe is equipped with the telescoping type of taper attachment.

10. Mark the approximate length of the taper on your workpiece.

12. Move the carriage to make sure that your tool clears the starting end of the taper to be cut by at least 3/4 in. If you don't have enough travel, adjust the location of the anchor bracket.

13. Select and set proper speed and feed.

14. Make a trial cut 1/16 deep and let it cut until it runs out of the work. Start the feed about 2 in. before the beginning of the cut to remove and play in the taper attachment.

15. Take a second lighter cut 0.020 in. deep so that a measurement can be taken.

16. With a pencil or scribe, make two lines exactly 1 in. apart on the tapered part of the workpiece.

17. Determine the correct taper per inch for your center's dimensions.

    TPI = taper per foot / 12   Record your results here: _____

18. Now, use a micrometer to measure your workpiece on the two lines. The difference between these diameters should be the same as the taper per inch of your center, as listed in the taper data tables.

19. If the taper is not correct, readjust the taper attachment. Then take trial cuts and measurements until taper dimensions are correct.

20. Take additional cuts until a smooth finishing cut makes dimension C correct. Surface finish should be very smooth.

21 Use a taper ring gage to check for an exact fir before getting to the final dimensions.

23. Remove the workpiece from the chuck and remove the chuck from the lathe.

24. Mount the workpiece in the lathe spindle nose. If necessary use a sleeve to match the size of the spindle nose taper..

25. Set the compound rest to 30 degrees.
26. Turn the 60 degree taper until the cylindrical part is 4 in. long.
27. This completes the turning of the lathe center. Save it to be used as a test plug for the sleeve and later to be turned in for grading.

EXERCISE 2. Turning an Internal Morse Taper Sleeve

Make a Morse taper sleeve to fit the live center.

Materials

1-1/4 x 2-5/8 in. HRMS for a No.3 Morse taper

Procedure

1. Select the diameter of internal taper that corresponds to the taper number of the lathe taper that have made.
2. Cut off material 1/8 in. longer than the workpiece (Figure 2).

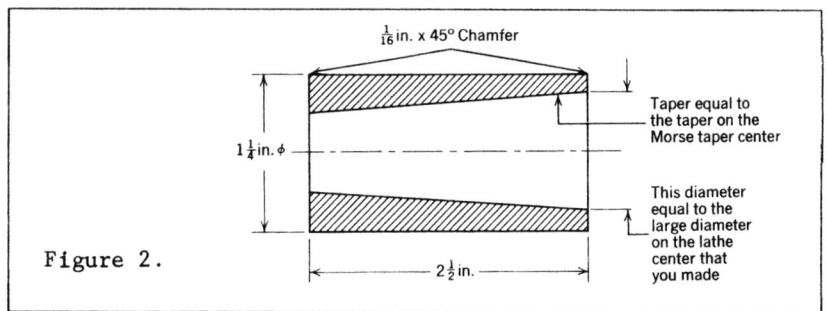

Figure 2.

3. Mount work in a three-jaw chuck. Face and drill the center.
4. Use a drill 1/16 in. smaller in diameter than the small diameter of the taper sleeve and drill completely through it.

5. Determine the taper per. foot for your taper. Record here:_____
6. Wipe the taper attachment clean and lubricate it.
7. Adjust the taper attachment to the required taper.
8. Mount a boring bar in your tool post; the cutting edge must be exactly on center. Check the tool to see that it has enough clearance.
9. Lock the anchor bracket to the lathe bed; move the carriage by hand to make sure you have sufficient travel.
10. Disengage the cross feed nut by removing the screw on the cross slide. Cover the hole to keep chips out. This procedure is not necessary if your lathe is equipped with the telescoping taper attachment.
11. Tighten the binder handle on the slotted extension of the cross slide.
12. Select and set your speed and feed.
13. Make trial cuts until the taper extends part of the way through the bore.
14. Use the lathe center that you made for a plug gage to check the internal taper. 15. On the last few passes use a fine feed of .003 to .004 in. for a smooth finish. 16. When the large diameter of the internal taper is the same diameter as the large diameter on the lathe center, the taper is finished. Chamfer the sharp edges.
17. Turn in the grading sheet and finished project to your instructor for evaluation.

PROJECT 10. INTERNAL THREAD

INSTRUCTOR EVALUATION

NAME_____DATE_____

Grade

	Letter	Percent
1. Follows drawing (dimensions and tolerances)	_____	_____
2. Machining finishes	_____	_____
3. Mechanism or tool operates satisfactorily	_____	_____
4. General workmanship	_____	_____
Total grade	_____	_____

Comments

Instructor signature: _____

PROJECT 10. INTERNAL THREADS

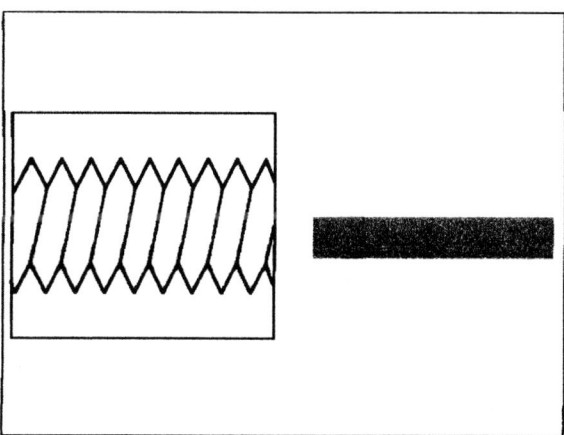

Objective

Learn how to produce an internal thread by single point lathe threading procedures.

Outline for Study

Prior to starting this project, study and complete the text unit, and Post-Test for:

1. Cutting an internal thread with a single point tool on the lathe in **Section I**

Procedures

Begin the procedure for this project by completing the following worksheet:

Worksheet 1. Cutting an Internal Thread on a Lathe

WORKSHEET 1. Cutting an Internal Thread

Tools and Material

One inch length of 1-1/2 in. dia. hexagonal or round stock

A precision 1-8 thread plug gage-, or a shop-made thread plug gage, or a standard 1-8 UNC cap screw

Procedure

1. Chuck the nut blank and make a facing cut. Chamfer. 2. Calculate the minor diameter for the 1-8 UNC thread. The formula is:

$$d = D - (P \times .541 \times 2)$$

where

P = pitch — 1/8 in. = .125 in.
D = 1 in. - .125 in. x .541 x 2
d = .86475 in. or .865

3. Spot drill the nut blank and drill through with a 3/4 to 13/16 in. drill

4. Set up a boring bar and tool. Bore the hole to .865 in., the minor diameter of the thread.

5. Set up your lathe for threading 8 TPI and place a threading tool in the boring bar.

6. Swivel the compound rest to 29 degrees toward the chuck.

7. Align the tool with the center gage. Check to see that the chuck jaws will clear the compound when the tool is moved through the bore.

8. Take the slack out of the compound screw by turning it outward and set the tool to the work. Zero both micrometer collars.

9. Take a scratch cut and check the number of threads with a screw pitch gage.

10. Select a slow speed for threading. Use cutting oil.

11. Determine the infeed on the compound for a 1-8 UNC Unified thread. The formula is:

    Infeed = P x .708   (P= pitch or 1/n where n is the number of thread per inch)

    $\quad\quad\quad$ = .125 in. x .708

    $\quad\quad\quad$ = .088 in.

12. Advance the compound before each cut .005 to .010 in. depending on the rigidity of your setup. Reduce the infeed to .002 in. or less near the last few cuts.

13. When the total infeed depth has been reached, clean out all the chips from the threads and try the thread plug gage for fit. If it will not go in, run the tool through one or two times without advancing the compound slide and try the gage again. If the gage still does not fit, it indicates that the tool form, flat, or alignment is not correct, or that something on your setup has slipped out of place, or that tool spring has not allowed a full depth of cut. Sometimes, several free cuts can be taken to enlarge the diameter. If you are not getting results, continue advancing the tool .002 in. per pass and check with the thread plug gage between each cut.

14. Chamfer the thread and reverse the workpiece in the chuck.

15. Face this side of the nut to length and chamfer both the thread and the outside diameter.

16. Clean up your project in solvent.

17. Turn in the grading sheet and the finished project to your instructor for evaluation.

PROJECT 11. MULTIPLE LEAD THREADS

INSTRUCTOR EVALUATION

NAME_____DATE_____

Grade

	Letter	Percent
1. Follows drawing (dimensions and tolerances)	_____	_____
2. Machining finishes	_____	_____
3. Mechanism or tool operates satisfactorily	_____	_____
4. General workmanship	_____	_____
Total grade	_____	_____

Comments

Instructor signature: _____

PROJECT 11. MULTIPLE LEAD THREADS

MULTIPLE LEAD THREADS

Objectives

1. Be able to produce a two-start external thread on a lathe. 2. Be able to produce a two-start internal thread on a lathe.

Outline for Study

Prior to starting each procedure for this project, study and complete Post-Tests for applicable units in:

1. Cutting the external two lead thread: **Section I**

2. Cutting the internal two lead thread: **Section I**

Procedures

Begin the procedures for this project by completing the following worksheets.

Worksheet 1. Cutting the External, Two Lead Thread

Worksheet 2. Cutting the Internal, Two Lead Thread

WORKSHEET 1. Cutting the External, Two Lead Thread

Materials

4 in. of 1 in. dia. HRMS round bar stock

Procedure

Make a two-start, double lead unified screw thread having a pitch of .100 in. on a 1 in. dia. rod.

1. Cut off about 4 in. of the round stock, face and center drill one end.

2. Chuck about 1/2 in. and place the dead center in the other end.

3. Take a skim cut to remove the scale. Take a micrometer measurement and record it. This will be the basis for your calculations for threads. <u>Note:</u> If CRMS is used, this step is not necessary.

4. Make an undercut about 3 in. from the tailstock end to terminate the thread. The depth is based on .100 in pitch.

5. Since the lead on a two-start thread is twice the pitch, it will be .200 in or 5 TPI. Set the quick-change gearbox to 5 T?I.

6. The two starts must be 180 degrees opposed, so you can use any of the several methods explained in your textbook. The most convenient, if your lathe is so designed, is to use the threading dial to make the division, using alternating positions on the dial. This should be tested with a scratch cut. This system will only produce two leads. Setting the compound parallel to the ways is a method in which there is no limit to the number of starts. However, the gibs must be tightened when you use that method.

7. Set up the lathe tool for threading and zero the micrometer collars. Take a scratch cut.

8. Advance the compound .100 in. (assuming you are using that method) and take a second scratch cut. All of the scratch marks should be equally spaced. The pitch of .100 in. (or 10 TPI on the gage) should be checked with a screw pitch gage.

9. Calculate the depth of the external thread in terms of the pitch, not the TPI set on the gearbox. The depth of cut on the cross slide is:

    P x .613 = .100 x .613 = .0613

    or, if you are feeding in with the compound set at 29 degrees, the depth of cut is:

    P x .708 = .100 x .708 = .0708

10. Feed in, taking about .005 in. per pass on one lead at a time, leaving about .010 to .015 in. for finishing.

11. Now alternately, finish both leads to depth. Note: If you are using the compound at 29 degrees for infeed, be sure to back it out and set it to zero before you begin the second cut.

WORKSHEET 2. Cutting the Internal, Two Lead Thread

Materials

1 inch length of 1-1/2 in. dia. round or hexagonal stock

Procedure

Make one two-start nut to fit the just completed external two-start thread.

1. Set up the nut blank in a three-jaw chuck and face both sides. Chamfer the edges with a lathe tool.

2. Calculate the minor diameter. The formula is:

    d = D - (P x .541 x 2)

    as explained in Project 10, "Internal Thread." If the major diameter of the external thread is slightly under one inch, because of the clean-up cut, use that figure in this formula for D.

3. Drill through the nut blank just under the calculated minor diameter. Bore to size.

4. Set up for internal threading and take the two scratch cuts as you did for the internal thread.

5. Cut each lead separately as you did with the external thread, leaving .010 in. for finishing.

6. When you are near finish depth, make cuts alternately from one lead to the other, checking the fit with the external thread between each pass. When the thread will turn in all the way in both positions (180 degrees opposite the start), the job has been completed.

7. Clean up both pieces in solvent.

8; Turn in the grading sheet and finished project for your instructor's evaluation.

PROJECT 12. ACME THREADS

INSTRUCTOR EVALUATION

NAME_____DATE_____

	Grade	
	Letter	Percent
1. Follows drawing (dimensions and tolerances)	_____	_____
2. Machining finishes	_____	_____
3. Mechanism or tool operates satisfactorily	_____	_____
4. General workmanship	_____	_____
Total grade	_____	_____

Comments

Instructor signature: _____

PROJECT 12. ACME THREAD

ACME THREAD

Objectives

1. Learn to cut external Acme threads on a lathe.

2. Learn to cut internal Acme threads on a lathe.

Outline for Study

Prior to starting each procedure for the project, study and complete Post-Tests for applicable units in:

Making the external Acme thread: **Section I**

Making the internal Acme thread: **Section I**

Procedures

Begin the procedures for this project by completing the following worksheets.
Worksheet 1. External Acme Threads

Worksheet 2. Internal Acme Threads

WORKSHEET 1. External Acme Threads

Material

1-1/2 in. piece of round HR or CR steel 5-1/8 in. long

Procedure

Make a thread plug gage (Figure 1) for a 1-3/8-4 Acme thread.

Figure 1.

1. Machine the plug gage blank as shown in Figure 1. Make the undercut 0.005 in. under thread depth.

2. Grind an Acme tool to the correct angles and dimensions as shown in your textbook. 3. Set up the lathe, for cutting threads.

4. Set up and align the tool to the workpiece.

5. Cut the Acme thread to the correct depth as explained in the textbook.

6. Show the finished plug gage to your instructor before continuing on. Save this gage for checking the Acme nut in Worksheet 2.

WORKSHEET 2. Internal Acme Threads

Material

Piece of hexagonal or round 2 in. diameter CF steel 1-1/4 in. long

Procedure

Make an internal 1-3/8-4 Acme thread in a nut to fit the Acme plug gage.

1. Face the nut blank to length and drill a hole slightly less than the minor diameter.

2. Determine the minor diameter and record it here: _____. Bore the hole to the minor diameter of the thread.

3. Set up the Acme tool in the boring bar and align it with a gage.

4. Make a scratch cut and check the pitch.

5. Continue threading until the tool has reached the major diameter.

6. Clean out the threads and check the fit with the plug gage.

7. Continue to take cuts of one or two thousandths of an inch until the required fit has been made.

8. Chamfer the nut.

9. Turn in the grading sheet and finished project to your instructor for evaluation.

PROJECT 13. CARBIDE TOOL EXERCISE

INSTRUCTOR EVALUATION

NAME_____DATE_____

Grade

	Letter	Percent
1. Follows drawing (dimensions and tolerances)	_____	_____
2. Machining finishes	_____	_____
3. Mechanism or tool operates satisfactorily	_____	_____
4. General workmanship	_____	_____
Total grade	_____	_____

Comments

Instructor signature: _____

PROJECT 13. CARBIDE TOOL EXERCISE

Objectives

1. Learn to select carbide inserts and tooling for specific requirements.

2. Use carbide tools on a lathe with various speeds, feeds, chip breakers and depths of cut in order to discern machinability.

3. Be able to form chips for better chip control

Outline for Study

Prior to starting the procedure for this project, study and complete Post-Tests for applicable units in: **Section F**

Procedures

Begin the procedures for this unit by completing the following worksheet.

Worksheet 1. Machinability Exercises
    Exercise 1. Tool and Chip Control

    Exercise 2. Speed and Its Influence on Cutting

    Exercise 3. Use of Formed Groove Chip Breakers

WORKSHEET 1 Machinability Exercise

Tools and Materials

    A 12 to 15 inch swing engine lathe and tooling

    One piece of HR steel round bar 2-1/2 to 3-1/2 in. diameter approximately 14 in. long

Procedure

Complete the exercises that show the influence of feed and speed, depth of cut and chip breaker geometry on chip control. Record your results.

EXERCISE 1. Feed and Chip Control

1. Mount a steel bar in a three-jaw or four-jaw chuck. Support the tailstock end with a live center.

2. Select a steel cutting grade carbide tool; having a separate adjustable chip breaker. Set a chip breaker width of 1/16 in. List your carbide identification and selection. _____

3. Mount the toolholder securely in a quick-change tool post with a minimum of overhang. The cutting edge must be at the lathe spindle centerline.

4. Select a cutting speed of approximately 300 to 40 SFPM. Determine RPM and record. RPM = _____

5. Set a feed of 0.002 in. per revolution.

6. Set the tool for a 0.060 in depth of cut.

7. Make a cut 1/2 t 1 in. long. Disengage the feed.

8. Record your finding on the Machinability Exercise Record Sheet. The chip will not be controlled, but should form as a continuous ribbon.

9. Increase the feed rate to 0.004 in. per revolution and make a cut no more than 1 in. long. Record your findings and repeat this feed increasing exercise until the chip is broken into figure 9 or letter C shaped short lengths.

10. Continue to increase the feed until the chip becomes too tight. This condition is shown as a linking and welding of the chip, or a very short and tight letter C configuration. More horsepower is required to produce these chips.

11. Change the chip breaker to a width of 3/32 in.

12. Starting with the same depth of cut and the speed as previously used, and a feed rate of 0.010in., take cuts as before and record the results after each cut. With the wide chip breaker, the chips should be controlled at heavier feeds.

EXERCISE 2. Speed and its Influence on Cutting Performance

1. Set up the lathe as for the first exercise.

2. The tool should have a 1/16 in. wide chip breaker.

3. Set the same feed that gave good chip control in the first exercise and a 0.060 in. depth of cut.

4. Set a speed of 100 SFPM and make a 1 in. long cut. Record the results. Fill in all the spaces on the record sheet.

5. Make successive cuts, increasing the speed by 50 SFPM for each cut until 400 SFPM is reached. Then, continue with 100 SFPM increments. Records your findings after each cut.

6. How did the speed changes affect the surface finish and the chip control ?

EXERCISE 3. Use of Formed Groove Chip Breakers

1. Set up the lathe as for the prior exercises.

2. Select a tool with a grooved chip breaker.

3. Set a .060 in. depth of cut.

4. Set a speed that gave good results in the previous exercise.

5. Start with a feed rate of 0.020 in. and make a cut. Record the results.

6. In crease the feed rate by 0.020 in. increments, taking cuts and recording your findings. Stop making feed increases when chips are getting too tight.

7. How effective are formed groove chip breakers ? Why ?

8. Make a chart as shown the next page on a full sheet of paper and record your results.

9. Turn in the grading sheet and the record sheet that you have prepared to your instructor for evaluation.

Machinability Exercise Record Sheet

			Work Material		Tool Identification		
FPM	**RPM**	**Feed**	**Depth of cut**	**Chip breaker width**	**Chip form form**	**Surface finish of work**	**Tool wear**

PROJECT 14. VEE BLOCKS

INSTRUCTOR EVALUATION

NAME_____DATE_____

Grade

	Letter	Percent
1. Follows drawing (dimensions and tolerances)	_____	_____
2. Machining finishes	_____	_____
3. Mechanism or tool operates satisfactorily	_____	_____
4. General workmanship	_____	_____
Total grade	_____	_____

Comments

Instructor signature: _____

PROJECT 14. VEE BLOCKS

Objectives

1. Use a horizontal mill to square a block and cut vees.
2. Learn heat treating processes for hardening and tempering alloy tool steel.
3. Learn to use Rockwell and Brinell Hardness Testers.
4. Use a precision surface grinder to square and finish vee blocks.
5. Learn to use precision measuring techniques on a surface plate.

Outline for Study

Prior to starting each procedure for this project, study applicable units and complete Post-Tests for:

1. Milling the blocks: **Section K**
2. Hardening the blocks: **Section D**
3. Grinding the blocks: **Section L**

Procedures

Begin the procedures for this project by completing the following worksheets.

Worksheet 1. Milling the Vee Blocks

Exercise 1. Indicating in the Machine Vise
Exercise 2. Surfacing a Workpiece after the Vise Has Been Indicated In
Exercise 3. Squaring a Workpiece in a Horizontal Mill
Exercise 4. Squaring the End of a Workpiece after the Sides Have Been Squared
Exercise 5. Machining a V-Shaped Part in the Horizontal Mill

Worksheet 2. Hardening and Tempering the Vee Blocks

Worksheet 3. Testing the Vee Blocks

Exercise 1. Rockwell Hardness Testing
Exercise 2. Brinell Hardness Testing

Worksheet 4. Surface Grinding the Vee Blocks

Exercise 1. Care of Gage Blocks and Wringing Practices

WORKSHEET 1. Milling the Vee Blocks

Material

3 inches of 2 x 2 in. SAE 4140 HR bar stock

Procedure

Mill two vee blocks as shown in Figure 1

Figure 1.

1. When you have cut off the 5 in. length of alloy steel, leave it full length for squaring operations. The blocks will be sawed in half later to make two blocks.

2. Deburr the material and prepare the horizontal milling machine as follows.

EXERCISE 1. Milling the Vee Blocks

The steps necessary to indicating-in (aligning) the milling machine vise. Note: This exercise is not needed if the mill vise you are using is aligned to the table with keys.

1. Attach a dial indicator to the arbor or overarm.
2. Position the indicator on the solid jaw of the vise.
3. Move the table by hand so the vise jaw moves past the indicator.
4. Adjust the position of the vise until the indicator readout is zero at each side of the vise jaw.
5. Follow the sequence for squaring a vise in **Section K** in the textbook.

EXERCISE 2. Indicating-In the Machine Vise

The steps necessary to surfacing the workpiece. Follow the sequence in **Section K** in the textbook.

1. Secure the workpiece in the vise.
2. Install a plain milling cutter (slab mill) that is wider than the workpiece.
3. Position the cutter.
4. Set the speed and table feed.
5. Turn on the cutting fluid and take the first roughing cut.

EXERCISE 3. Squaring a Workpiece in a Horizontal Mill

The steps necessary for squaring a workpiece in a milling machine after one flat surface has been obtained. Follow the sequence for squaring a workpiece in **Section K** in the textbook.

1. Deburr the workpiece and clean the vise so it will have an accurate reference surface.

2. Place the deburred flat surface against the solid jaw.

3. Approximately equal amounts should be taken on all sides since hot rolled bars tend to have a decarburized "skin" which is usually about 1/32 in. deep. Decarburization causes the surface to remain soft after the part is hardened.

4. Be careful <u>not</u> to cut to the dimensions given in Figure 1. Those are finished dimensions which will later be obtained by precision grinding. You must leave about 0.030 in. overall for finishing, 0.015 in. on each side. Slight errors in the milling operation (up to 0.003 in. out of square or taper) are permissible since the grinding operation will correct these errors. If the error is greater than this amount, it may still be possible to "save" the piece. Ask your instructor.

5. Remove the part, deburr, and clean the vise and parallels.

6. Check the part with a square and check all of the dimensions with a micrometer.

EXERCISE 4. Squaring the End of a Workpiece after the Sides Have Been Squared

The steps necessary to squaring the ends of a workpiece after the sides have been squared. Follow these sequences for milling the ends of a workpiece in **Section K**, in the textbook.

1. Place the 5 inch long block in the cutoff saw and cut it into two equal halves. This leaves enough material on each block for milling the ends square.

2. Place the workpiece in the milling vise with the end up and align it with a square.

3. Take a clean-up cut with the slab mill.

4 Reverse the workpiece and mill the opposite end to 0.030 in. over finish size.

5. Remove the workpiece and deburr.

EXERCISE 5. Machining a Vee-Shaped Part in the Horizontal Mill

The steps necessary to machine a vee-shaped part in the horizontal milling machine.

1. Position the mill vise so one block can be gripped on the ends while supported in a vee-block that is shorter than the workpiece (Figure 2). Make sure the workpiece is well seated against the vee-block. A 90-degree vee can be milled in this manner.

2. Mount a large side-cutting milling cutter on the arbor. The cutter must be large enough to cut to the layout line without allowing the arbor to touch the workpiece.

3. Cut to the layout line of the vee (Figure 3). Several passes should be used since this is not a good setup for heavy cuts. In conventional milling, the cutter could possibly lift the workpiece out of the vise if a full depth cut were made.

4. Repeat the procedure on the second block.

5. Mill the smaller vees on the opposite sides of both blocks.

6. Change to a 1/2 inch wide side-cutting cutter and mill the 1/2 inch slots on the sides. These may be cut in one pass in a No. 2 or larger horizontal mill. The blocks may be held in the vise on parallels in much the same way as when the vees were cut. Place the blocks as far down as possible into the vise jaws.

Figure 2. 90-degree vee can be milled with this setup

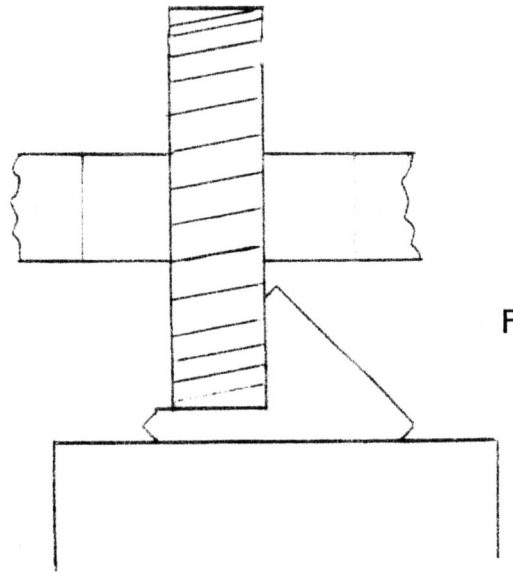

Figure 3. A side-cutting milling cutter is required for this operation

7. Slitting saw 3/16 and 1/8 in. wide should be used to cu the slots in the bottom of the vees.

8. All edges should be evenly chamfered with a file as shown on the vee block drawing.

9. A slot should be milled 3/8 in. wide and 1/16 deep in both blocks so that a name or initials can be stamped on it for identification purposes.

10. Machine and floor areas should be cleaned and tooling put away.

WORKSHEET 2. Hardening and Tempering the Vee Blocks

Materials

    Two heat treating furnaces

    Oil quenching bath

    Tongs and safety equipment

    Previously machined set of SAE 4140 steel vee blocks

Procedure

    (a) Harden the vee blocks

    (b) Temper the vee blocks to Rc 48 to 52 (HB 470 to 514).

Note: Soaking times for hardening and tempering operations are based on the smallest cross section of the part. About 3/4 to 1 in. may be considered as the least cross section of this vee block.

1. Set the furnace thermocouple control for the correct hardening temperature for the SAE 4140 steel, which is 1550°F (843°C).

2. Set a second furnace to the correct tempering temperature and turn it on. Consult the SAE 4140 mechanical properties chart (Figure 4) to determine the temperature to obtain a draw temper hardness of Rc 48 to 52 (HB 470 to 514). Note on Figure 4 the 4140 is HB 601 (approximately Rc 60) in the as-quenched condition. In order to reduce this hardness to about HB 500, the temperature would need to be set at about 550°F (288°C) on the tempering furnace. See Table D-1 in the textbook.

3. The vee blocks should be protected from decarburization by any of several means. A simple method is to first warm the blocks to several hundred degrees (about 400°F), sufficient to melt the boric acid powder. Lightly sprinkle boric acid powder on the hot vee blocks on all sides and place them in the furnace on a fire brick; use tongs.

4. After the vee blocks are the same color as the furnace, allow a soaking time of 1 hour per inch of least cross section (about 3/4 to 1 in. for 3/4 to 1 hour).

5. Using face shield and gloves, heat the tongs on the gripping end. Remove one vee block from the furnace, close the door, and quickly plunge the block into the oil bath, agitating the block until it has cooled. It should still be warm to the touch. Repeat the process with the second vee block

6. Place the two warm vee blocks immediately into the tempering furnace and hold them at that temperature for 1/2 hour.

7. Remove the blocks and allow them to cool in air.

8. Check for hardness. Because of possible decarburization, the readings may be low. Check again after surface grinding. See Worksheet 3 for hardness testing.

9. Is your vee block the same hardness that the mechanical properties chart indicated it would be at your selected tempering temperature? If not, what reason can you give for the difference?

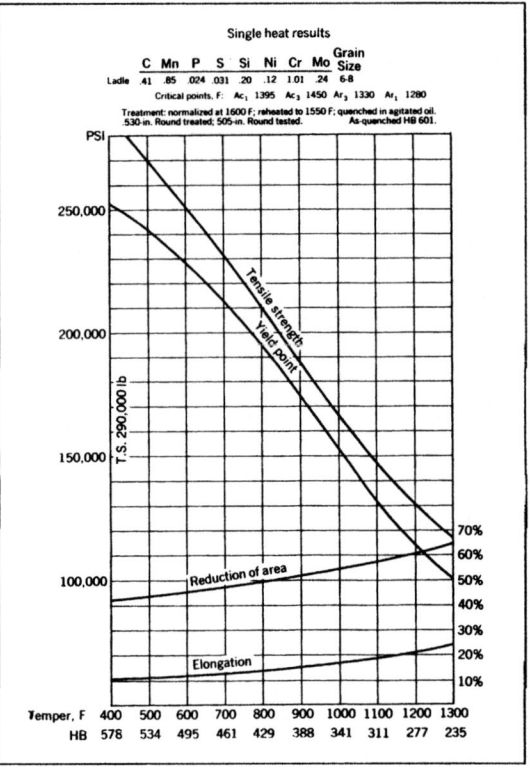

Figure 4. Mechanical Properties chart for SAE 4140 steel (Courtesy of Bethl

WORKSHEET 3. Testing the Vee Blocks

EXERCISE 1. Rockwell Hardness Testing

Materials

Rockwell hardness tester
Hardened steel
18 gage sheet steel

Nonferrous metal
Tungsten carbide insert
Small part with nitrided steel case

Procedure

(a) Three hardness tests on the standard Rockwell tester.
(b) Two tests on the superficial tester.

1. Before starting, see that the crank handle is forward.
2. Select proper Penetrator and insert it in the plunger rod.
3. Place the proper anvil on the elevating screw.
4. Select the proper weights.
5. Place the specimen on the anvil.

1. Place the specimen on the anvil.
2. Raise the specimen until it touches the ball.
3. Apply the load for 30 seconds. Release.
4. Remove the specimen from the tester and measure the diameter of the impression.
5. Determine the Brinell Hardness Number.
6. Check for hardness in two or three locations on the specimen.

	Material	Surface	Weight	Reading
1.	Nonferrous metal			
2.	Hardened steel			
3.	Soft steel			

Record your results and show them to your instructor.

Testing the Blocks

1. After completing the Rockwell and Brinell hardness testing exercises (on both or whichever tester is available to you), you are now ready to test your vee block.

2. Take three tests in different locations and record the results. If the tests do not average 48 to 52 RC, try again after some of the surface has been removed by surface grinding. If the blocks did not get hard for some reason, they may be rehardened; but first they must be normalized as explained in **Section D**

WORKSHEET 4. Surface Grinding the Vee Blocks

Tools and Materials

Surface grinder and accessories
A set of vee blocks previously machined and heat treated

Procedure

1. Grind all sides square to each other within plus or minus 0.0005 in.

2. Hold the decimal dimensions to within 0.0005 in. and the fractional dimensions within plus or minus 1/64 in. Follow the steps given in **Section L** in your textbook.

3. Obtain a fine finish (about 16 to 20 microinch) on all surfaces with the final grinds. Check with your instructor on measurement, squareness, and finish from time to time as you progress with this project.

6. Raise the specimen into contact with the penetrator by turning the capstan handwheel clockwise. Continue motion until the small pointer is near the dot. Continue until the larger pointer is in a vertical position. The minor load is now applied.
7. Turn the bezel of the dial gage until the "SET" line is directly behind the large pointer.
8. Release the weights (major load) by tripping the crank handle clockwise. Do not force the crank. Allow it to come to rest.
9. When the large pointer has come to rest, return the crank handle to the starting position. This removes the major load; the minor load is still applied.
10. Read the scale letter and Rockwell hardness number from the dial gage.
11. Remove the minor load by turning the capstan hand wheel counterclockwise to lower the elevating screw and specimen.
12. Remove the specimen. Repeat the test procedure in one or more locations on the test piece.

Material	Surface Condition	Rockwell Penetrator Scale	Weight	Hardness Reading 1 2 3 Average
1. Nonferrous metal				
2. Hardened steel				
3. Tungsten carbide insert				
4. 18 gage sheet steel				
5. nitrided steel part				

Record your results and show then to your instructor.

Brinell Hardness Testing

Materials

Brinell hardness tester
Heat treated steel
Nonferrous material (aluminum or copper)
Soft steel

Procedure

(a) Make three hardness tests.

EXERCISE 1. Care of Gage Blocks and Wringing Practices   **107**

Tools and Materials

Shop set of gage blocks
Clean solvent
Lint free tissue Camel's
hair brush Gage block tray
Dial test indicator

Height gage
Surface plate
Insulated forceps (if available)
Shop cloth or lint free cloth
Gage block preservative
Conditioning stone (granite or ceramic) (Be sure that this is the correct item!)

Procedure

1. Select two gage blocks from each size series and lay them on the tray which has been covered with a clean shop towel or lint free cloth. Select two wear blocks.

2. Moisten a lint free tissue in clean solvent and clean the surface of the conditioning stone. Also clean the surfaces of the gage blocks to be wrung.

3. Dry the surfaces of the conditioning stone and the blocks to be wrung.

4. After cleaning, place the blocks on their sides to reduce the amount of dust settling on the cleaned surfaces.

5. If more than a few seconds elapse after cleaning and before wringing, dust the surface to be wrung with the camel's hair brush.

6. Bring the surfaces into contact and wring the blocks using the procedure described in the textbook, **Section C, Using Gage Blocks.**

7. If the conditioning stone must be used, first dust the stone with the camel's hair brush. Deburr the block with a single back and forth pass over the stone.

8. Wring the remaining blocks together, using a wear block at each end of the stack.

9. Make a stack 2k in. high to check the length of the vee blocks.

10. Set up a 0.0001 in. discriminating dial test indicator on a height gage and place them on a surface plate. Compare the height of the vee blocks to the gage block stack.

11. Repeat the process in steps 9 and 10 to check the 2 inch dimensions of the vee blocks.

12. Disassemble the blocks by sliding them apart. Spread them apart on the clean shop towel or tissue.

13. If you do not use gage blocks for measuring the vee blocks, disregard steps 9 through 11.

14. Turn in the grading sheet and the completed vee blocks to your instructor for evaluation.

PROJECT 15. PRECISION VISE

INSTRUCTOR EVALUATION

NAME_____DATE_____

Grade

	Letter	Percent
1. Follows drawing (dimensions and tolerances)	_____	_____
2. Machining finishes	_____	_____
3. Mechanism or tool operates satisfactorily	_____	_____
4. General workmanship	_____	_____
Total grade	_____	_____

Comments

Instructor signature: _____

PROJECT 15. PRECISION VISE

PRECISION VISE

Objectives

1. Learn to square a block and to machine steps and grooves in the block on a horizontal mill.
2. Be able to drill and ream holes and machine slots on a vertical mill.
3. Be able to precision grind for perpendicularity and dimension.

Outline for Study

Prior to starting each procedure for this project, study and complete Post-Tests for applicable units in:

1. Machining the block: **Section K.**
2. Milling the slot and cavity in the vise body: **Section J.**
3. Making the movable jaw and parts: **Section J**
4. Hardening and tempering the vise parts: **Section D.**
5. Grinding the vise parts: **Section L.**

Procedures

Begin the procedures for this project by completing the following worksheets.

Worksheet 1. Machining the Block for the Vise Body

Worksheet 2. Machining Holes in the Vise Body

Worksheet 3. Milling the Slot and Cavity

Worksheet 4. Making the Movable Jaw and Parts

Worksheet 5. Hardening and Tempering the Vise

Worksheet 6. Grinding the Vise Parts

WORKSHEET 1. Machining the Block for the Vise Body

Materials

6-1/4 inches of 2-1/2 x 3-1/4 in. SAE 4140 steel

Procedure

1. Set up the block in a horizontal milling machine as explained in **Section K**, in your textbook and mill it square and to the oversize (to allow for grinding) dimensions shown in Figure 1.

Figure 1.

2. Take the steel block that has been squared and machined to length. Make a layout of the grooves and the step to be machined. See Drawings I and II at the end of this project. Follow the procedures for side milling in Section K of your textbook.

3. Mount a vise on the milling machine table and align the solid jaw so it is square to the column face.

4. To machine the large step on the project, mount the workpiece in the vise. Support the workpiece on parallels so that about 1-5/16 in. extends above the vise top. The step is to be 1-1/4 in. deep; the extra 1/16 in. is to clear the vise jaws.

5. Select a side milling cutter with a diameter large enough to make a cut 14 in. deep.

6. Mount the cutter on the arbor so that the cutting pressure will be against the solid jaw. Check for sufficient clearance between arbor support and vise. Check the position of the cutter on the arbor by moving the table cross feed to see if the full width of the cut can be made.

7. Select and set the cutter speed. Use the formula

    RPM 4CS/D

8. Check that the spindle rotation is correct.

9. Calculate the feed rate to use and set it on the machine. The feed rate equals feed per **tooth x number of teeth x RPM.**

10. Set the depth of cut. One or two roughing cuts may be required and a finishing cut will also be needed. The final roughing cut should leave about .030 in. for the finish cut. Note that on Figure 1, the finish dimension is .030 in. oversize.

11. Move the table to set the width of the cut to 1/2 in. The width of the cut should be less than the width of the cutter.

12. Tighten the knee and cross slide locking clamps.

13. Turn on the coolant.

14. Start the cut manually, then engage the power feed (Figure 2). Observe the cutting operation carefully and be prepared to disengage the power feed at the first sign of trouble.

15. When the cut is completed, return the table to its starting position.

16. Move the table into position to take another 1/2 in. wide cut.

17. Repeat taking cuts until you are within 1/4 in. of your layout lines. Then stop the spindle and measure the width of the remaining section. Calculate the amount of oversize. Adjust the machine table by the amount of oversize using the cross feed micrometer collar.

18. Make this cut. Set the cross feed micrometer collar to zero. This is also the reference point for the finishing cut. This completes the roughing cut.

Figure 2.

Figure 3.

19. Return the machine table to the starting point for the first cut. Set the depth of cut for the finishing cut.
20. Make the finishing cuts in the same manner as the roughing cuts, except that you <u>stop the cutter</u> before returning it over the finished surface. Returning a revolving cutter over a surface that was just machined will leave cutter marks.
21. After completing the final cut, stop the machine. Remove the chips and deburr the workpiece. Measure the workpiece while it is still clamped in the vise. If another cut is needed, it should be made next. This completes the milling of the step. Remove the workpiece from the vise.
22. Turn the vise and align the solid jaw parallel with the table.
23. Clamp the workpiece with the long outside (bottom) surface against the solid jaw of the vise as shown in Figure 3.
24. Select a sharp side milling cutter to machine the grooves. Exchange the cutter on the arbor with the new cutter.
25. Position the cutter on the side of the workpiece by using a paper strip. Zero the cross feed micrometer collar.
26. Move the cutter over the groove location using the micrometer collar. The cutter should now be aligned with the layout lines.
27. Set the depth of cut. Only one cut is required for making the full groove depth.
28. Calculate and set the speed for this cut.
29. Calculate and set the feed rate.
30. Turn on the coolant and start the cut manually, then engage the power feed.
31. At the completion of this cut, turn off the feed, coolant, and spindle, and return the table to its starting position.
32. Move the table the required distance to align the cutter for the second groove. Use the micrometer collars to adjust the machine to get accurate locations.
33. Cut the second groove.
34. Stop the machine, then return the table to its starting position.
35. Clean off the chips, deburr the just machined grooves, and remove the workpiece from the vise.
36. Clean the vise and workpiece. Reposition the workpiece in the vise with the grooved side down and with the same aide against the solid jaw as used for the last operation. Clamp the workpiece securely.

37. Positioning of the workpiece under the cutter should be done with the aid of the micrometer adjustments. The grooves on the top side should be exact duplicates of the grooves on the bottom side. Remember to remove the backlash that exists when you change the rotation of any of the feed screws.
38. Cut both grooves.
39. The undercut slot on the jaw for grinding relief should now be cut, using a $\frac{1}{8}$ in. slitting saw. A cutter with a radius would be preferred as a round notch is less likely to crack during heat treatment operations.
40. Clean and deburr the workpiece and remove it from the vise. Clean the machine.
41. This completes the horizontal milling machine work on this workpiece.

WORKSHEET 2. Machining Holes in the Vise Body

Materials

Vertical milling machine and accessories

Procedure

1. Lay out the vise body you have just milled to the oversize dimensions shown in Figure 4.

Figure 4. Vise body holes

2. Clamp the workpiece in the vise on a vertical mill table with a side up. Set up a drill chuck in the spindle.

3. Using an edge finder, locate the spindle to the large end of the vise body and center it to the groove in which the holes are to be drilled.

4. Determine the distance to the first hole from Drawing II and add .015 in. for oversize. Do not forget to subtract one-half the diameter of the edge finder

5. Spot drill, drill under the reamer size, and ream one hole at each location. Then move to the next hole. Drill and ream only halfway through the block. See Figures 5 and 6.

6. When one side has been finished, turn the block over and set up again, and repeat the previous steps.

7. Now turn the vise body on end as shown in Figure 7 and drill and counterbore as shown in Drawing II.

WORKSHEET 3. Milling the Slot and Cavity

Materials

Vertical milling machine and accessories

Procedure

1. Fasten the workpiece in the vise bottom up; use parallels so the workpiece extends above the vise for measuring while it is being machined.
2. Use a 7/16 in. diameter center cutting end mill to rough out the 1/2 in. wide slot (Figure 8).

Figure 8. Vise Body

3. Calculate the RPM.
4. Calculate the federate.
5. Align the end mill over the slot and lock all machine axes that will not have any movement during the cutting operation.
6. Rough out the slot; use coolant while cutting.
7. Change to a 1/2 in. end mill.
8. Make a 1/8 in. deep trial cut, then measure the location of the slot in the workpiece. Make any necessary adjustments to put the slot into the center of the workpiece.
9. Finish machine the 1/2 in. slot.
10. Rough out the cavity (Figure 9), leaving 1/32 in. for the finish cut all around.
    Keep the cutting pressure against the tool by using conventional (up milling) techniques.
11. Measure the depth of the cavity with a depth micrometer. Then make the depth adjustment necessary for the finish cut.
12. Make a cut with the cutter just touching the side of the cavity, measure the distance with a micrometer, then adjust for one final cut the correct distance in from the outside edge.
13. Machine each side of the cavity, repeating the operations performed in step 12.
14. Deburr all edges.
15. This completes the machining of the vise body prior to heat treating and finish grinding.

WORKSHEET 4. Making the Movable Jaw and Parts

Materials

3-1/4 inches of 1-1/2 x 2 in. SAE 4140 HR steel

6-1/2 inches of 5/16 x 1-3/4 in. spring steel or SAE 4140 steel

2 inches of 1 in. dia. SAE 4140 steel round

Procedure for the Movable Jaw

1. Square the block in either a horizontal mill or by using a carbide face mill in a vertical milling machine. Leave it 0.030 in. oversize.
2. Clean and deburr the block and lay out for drilling and milling as specified in Drawing III.
3. Drill and tap the 1/4-28 UNF screw holes.
4. Set the milling head to 45 degrees and mill the angular surface as shown in Figure 10.
5. With the same setup, drill the 3/8 in. hole.
6. Set up a 1 in. dia. ball end cutter and machine the ball socket.
7. Turn the part in the vise so the bottom side is up and mill the recess. Remember to make it 0.030 in. less in width than the finish dimension.
8. Make the undercuts as shown on Drawing III.
9. Using a 3/8 in. end mill with the same setup, elongate the 3/8 in. drilled hole to make a slot that will be vertical to meet the ball socket on one end, and angular on the other. This slot will allow the 3/8 in. clamping screw free movement so it can clamp at different angles.
10. This completes the milling on the movable jaw. It can now be hardened and tempered along with the vise body.

Procedure for the Jaw Insert

1. Saw off 6-1/2 inches of the tool steel for the jaw inserts and mill them to the dimension shown in Drawing IV on the edges, allowing 0.015 in. for grinding. If chrome spring steel is used, remember it is prone to work hardening so the cutting speed must be very low.
2. Cut the piece in two, and mill the ends, leaving 0.030 in. oversize.
3. Lay out for the holes. Drill and countersink on one and drill and tap on the other.
4. The jaw inserts may be heat treated along with the vise body and movable jaw with the same settings for hardening and tempering.

Procedure for the Swivel Nut

1. The nut may be turned on a lathe as shown in Drawing IV.
2. Place the turned piece in the vertical mill and cut the angles.
3. The nut may be hardened along with the other parts, but it should be tempered at about 650°F (343°C) so that it will be tougher causing it to resist bending or breaking.

Procedure for the Ball Washer

1. The ball washer can be turned on a lathe by first drilling and counterboring the square end. Make a facing cut. Chamfer.
2. Use a ball turning attachment or free hand the radius toward the chuck, leaving a stem at least 3/8 in. diameter. Check frequently with a radius gage. Finish with a file and abrasive cloth.
3. The ball washer need not be hardened, but left in the as-rolled condition. A 5/16-24 UNF by 2 in socket head cap screw will also be needed to complete the jaw parts.

WORKSHEET 5. Hardening and Tempering the Vise

Materials

Heat treating furnaces and accessories; protective clothing, gloves, face protection Procedure

1. Hardening and tempering will be done in much the same way as that for the vee blocks and according to methods explained in Section D. Two furnaces should be used, one for hardening and one for tempering, so no delay will take place between the two operations.
2. Some means such as using a light coating of boric acid powder on the parts should be used to prevent decarburization and scaling.
3. The large parts should be put in a cold furnace and brought up to the hardening temperature. The small parts can be put in a hot furnace.
4. Holes in the larger parts should be plugged with steel wool, and a wire fastened in a convenient location so the part can be easily picked up in the correct orientation for quenching in the oil.
5. The quenching oil should be slightly warm to the touch. Note: A quench crack can instantly ruin a part that has many hours of work in it. Quench cracks are caused most often by a hardening temperature that is set too high. They can be caused at the intersection of a large mass and a small mass if they are not quenched correctly. In this case, arrange for the large end of the vise body to enter the quench bath first; do not hesitate to plunge it quickly under the oil and agitate it. If the quench bath is quite cold, the shock may cause a crack. For most small parts, a cold quench is acceptable, but with more massive pieces, the quench bath should be warmed. These parts seldom develop a quench crack and, when they do, it can always be traced to an extreme violation of good practice.
6. As soon as the parts have been quenched and cooled so they are still warm to the touch, they should be wiped off (to remove oil) and put directly in the tempering furnace and kept there for about one-half hour.
7. When the parts have been cooled, they should be tested for hardness.

WORKSHEET 6. Grinding the Vise Parts

Materials

Surface grinder and accessories; surface plate and precision measuring devices Procedure

1. Methods of surface grinding as explained in applicable units in Section L, will be used to grind the vise parts. The procedures will be much the same as those used for the vee blcks.

2. The bottom side of the vise body should be first ground by supporting the body on magnetic parallels.

3. Clean the part and wipe off the magnetic table. Place the vise right side up on the table, crosswise to the travel so the jaw can be ground in the same setting. The vise must be set square to the travel.

4. Grind the horizontal surface with a straight wheel. Change to a dish wheel and clean up the jaw face.

5. Replace the straight wheel and clean up grind all sides, leaving them oversize.

6. Use the squaring technique with a precision angle plate, surface gage, and dial indicator that you used on the vee blocks.

7. When the vise body and movable jaw are relatively square, check them for squareness on a surface plate with a precision cylindrical square. They should be within plus or minus .0005 in perpendicularity. Of course, they must still be oversize. Have your instructor observe this check.

8. When they are square within the acceptable tolerance, finish grind the grooves on the side of the body with a 1/4 inch wide straight wheel.

9. True up a straight wheel for finishing. Finish grind the vise body and the movable jaw to approximately 16 micro inches and to a dimensional tolerance of plus or minus .0005 in. Note: The movable jaw must have a sliding fit on the vise body; not over 0.003 in. clearance, but free to move.

10. The jaw inserts may now be surface ground to size, and the notches and chamfers ground.

11. ALL sharp corners and edges must now be slightly chamfered with a fine stone by hand. These sharp edges can cause a severe cut unless they are dulled.

12. Clean and oil all parts. You may now assemble the precision vise.

13. Turn in the grading sheet and the completed precision vise to your instructor for evaluation.

Figure 5. Drilling 11/32 in. diameter holes in the vise body (Lane College)

Figure 6. Reaming 3/8 in. diameter holes in the vise body (Lane College)

Figure 7. Drill and counterbore holes in the solid jaw of the vise body (Lane College)

Figure 9. Milling out the cavity

Figure 10. Using a shell mill to machine the angle

PRECISION VISE: DRAWING I

All sides square to each other to within 0.0005 in. Dimension tolerance ± 0.0005 for the decimal dimensions ± 1/64 in. for fractional dimensions. Movable jaw sliding clearance is 0.015 to 2.003 in.

DET	PART NAME	REQD
9	NUT	1
8	JAW	1
7	BALL WASHER	1
6	5/16 x 1 1/2 NF SOC HD CAP SCR	1
5	1/4 x 1 FLAT HD MACH SCR	2
4	JAW INSERT	1
3	JAW INSERT	1
2	1/4 x 3/4 SOCKET HD CAP SCR	2
1	BASE	1

PRECISION VISE SCALE: FULL PLATE NO. 1
MACHINE TOOLS AND MACHINING PRACTICES

PRECISION VISE: DRAWING II

PRECISION VISE: DRAWING III

PRECISION VISE: DRAWING IV

PROJECT 16. CYLINDRICAL GRINDING

INSTRUCTOR EVALUATION

NAME_____DATE_____

	Grade	
	Letter	Percent
1. Follows drawing (dimensions and tolerances)	_____	_____
2. Machining finishes	_____	_____
3. Mechanism or tool operates satisfactorily	_____	_____
4. General workmanship	_____	_____
Total grade	_____	_____

Comments

Instructor signature: _____

PROJECT 16. CYLINDRICAL GRINDING

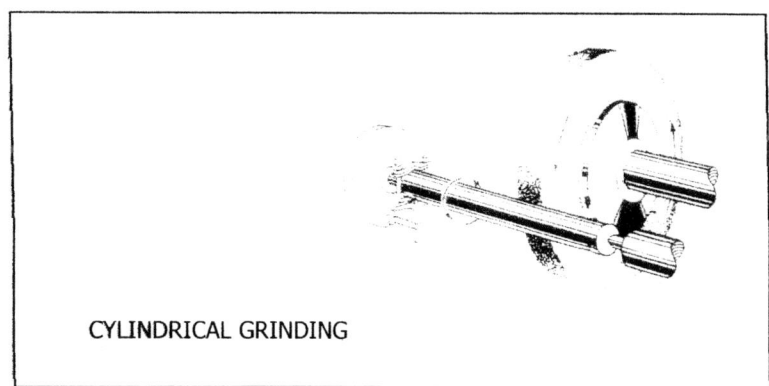

CYLINDRICAL GRINDING

Objectives

1. Learn to use the cylindrical grinding machine to make a straight precision cylinder.
2. Be able to produce a precision taper.
3. Be able to grind an internal surface.

Outline for Study

Prior to starting each procedure for this project, study and complete Post-Tests for:

1. Grinding a cylindrical shape: Applicable units in **Section L**.
2. Grinding external and internal tapers: **Section I**

Procedures

Begin the procedures for this project by completing the following worksheets.

Worksheet 1. Making a Lathe Mandrel: Grinding a Cylinder
Worksheet 2. Grinding a Taper
Worksheet 3. Grinding a Bore

WORKSHEET 1. Making a Lathe Mandrel; Grinding a Cylinder

Material

7-1/8 in. of 1 in. SAE 4140 steel for a 3/4 in. mandrel or sufficient material for a

selected mandrel size. Standard mandrel dimensions may be found in Machinery's Handbook.

Procedure

1. Turn a mandrel between centers on a lathe as shown in Figure 1. Check with your instructor to see if a different size or another part is to be made.
2. Make the turned diameter 0.015 to 0.030 in. oversize (depending on size).
3. Mill the flats on the ends.
4. Harden and temper to about RC 50.

Figure 1. Tapered Lathe Mandrel

5. Follow the instructions in the applicable unit in section L of your textbook, for grinding a mandrel. Before beginning the mandrel grinding procedure, make a straight grind.
6. When you have made the setup, before turning on the wheel, have your instructor inspect it.
7. With the wheel and work rotating and the table traverse set, feed the wheel to the work. Infeed the wheel until it just begins to produce sparks.
8. Turn on the coolant.
9. Make several passes, in feeding about 0.001 to 0.002 in. per pass until the surface scale is all cleaned up. Allow the table to traverse without infeed until it sparks out. Retract the wheel and stop the rotation and table traverse.
10. Check the two ends of the mandrel with a vernier or "tenth" micrometer. It should be the same within one ten thousandth (.0001) in.
11. If there is taper, move the table to compensate and regrind until the mandrel is all the same diameter within plus or minus 0.0001 in. The mandrel should still be over the final finish diameter.

WORKSHEET 2. Grinding a Taper

Materials

The mandrel

Procedure

1. Follow the steps given in your textbook on page 689 for grinding a taper. When adjusting the swivel table for the required taper, a dial indicator may be used in a position near the end of the table at a given distance from the pivot point of the table. This distance and the dial indicator reading is a proportional ratio to the ratio of amount to be removed from the mandrel <u>on one side</u> in a given length. For example, in 6 inches the mandrel should taper 0.003 in. in diameter. The table must be rotated so that 0.0015 in. more is removed from one end <u>on one side</u> than is removed on the other. If the table pivot point is 30 inches from the dial indicator, the ratio would be 30 in. to 6 in. or 5:1. Therefore, the dial indicator movement should be 5 x 0.0015 = 0.0075 in.
2. The final dimension of the taper should be determined by making the nominal diameter about one-fourth of taper length away from the large end. For example,

    the 3/4 inch mandrel should be 0.750 in at 1-3/16 in.
3. When you are near the final size, you should have an acceptable finish. If you do not, check with your instructor.
4. Clean up your finished mandrel.
5. Save the mandrel to turn in with the Worksheet 3 project.

WORKSHEET 3. Grinding a Bore

Materials

Any part or project requiring an internal ground surface.

Note: Your instructor may or may not wish you to do this exercise. This exercise can also be done on a tool and cutter grinder that has a cylindrical grinding attachment.

Procedure

1. Whether you use the high speed attachment on the cylindrical grinder or the cylindrical grinding attachment on a tool and cutter grinder, the method is the same.

2. The part to be internally ground is usually mounted in a precision chuck and set up to run true. The rotational speed will need to be adjusted for optimum finish and so that the work surface and wheel surface are moving in opposite directions.

3. A small diameter mounted wheel is usually used for small bore internal grinding. The shank of the wheel is held in a collet in the spindle. The wheel speed will need to be set according to diameter and within safe wheel speed. Table movement, wheel dressing, and use of coolant is similar to external grinding.

4. Use light cuts to produce good finishes. Grind to diameter.

5. Clean up the workpiece and the equipment you used.

6. Turn in the grading sheet, the mandrel, and the completed internal grinding project to your instructor for evaluation.

PROJECT 17. TOOL and CUTTER GRINDING

INSTRUCTOR EVALUATION

NAME_____DATE_____

Grade

	Letter	Percent
1. Follows drawing (dimensions and tolerances)	_____	_____
2. Machining finishes	_____	_____
3. Mechanism or tool operates satisfactorily	_____	_____
4. General workmanship	_____	_____
Total grade	_____	_____

Comments

Instructor signature: _____

PROJECT 17. TOOL and CUTTER GRINDING

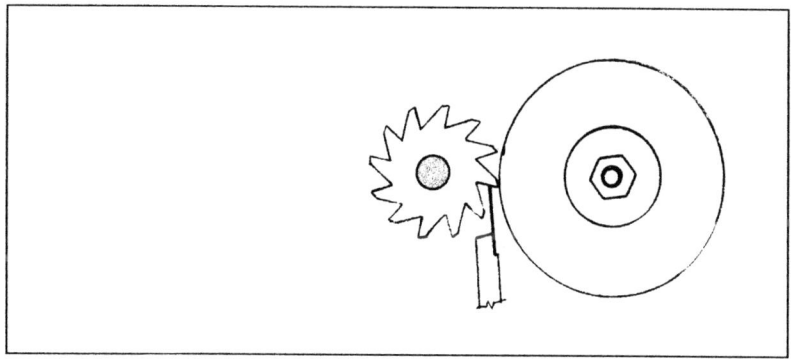

Objectives

1. Learn to sharpen plain and slab milling cutters, and also side cutting types.
2. Learn to sharpen end mills on the end and periphery.

Outline for Study

Prior to starting each procedure for this project, study:

1. Sharpening milling cutters: **Section L**

Procedures

Begin the procedures for this project by completing the following worksheets.

Worksheet 1. Grinding a Plain Helical Milling Cutter
Worksheet 2. Sharpening a Side Cutting cutter
Worksheet 3. Sharpening an End Mill

WORKSHEET 1. Grinding a Plain Helical Milling Cutter

Materials

Universal tool and cutter grinder and attachments One or more dulled helical milling cutters

Procedure

1. Mount the cutter on a grinding arbor as shown in your textbook **Section L**
2. Set up for grinding the primary clearance as explained on pages 698 to 700 in your textbook.
3. Do not attempt to grind below severe gouges and nicks in the cutter teeth. It is a more economical procedure to just remove dullness and damage that is not more than 0.005 to 0.015 in. deep. If there is a question on how much to remove, ask your instructor.
4. Show your sharpened cutter to your instructor.

WORKSHEET 2. Sharpening a Side Milling Cutter

Material

One or more side milling cutters, such as staggered tooth types

Procedure

1. If you are sharpening a staggered tooth cutter, set up a toothrest like that shown in Section L. If the teeth on the cutter are widely spaced, use a toothrest like the one on the upper right in Figure L-13; but if the teeth are close together, use one like the one below it, that is, an offset toothrest.
2. The wheel should be a narrow straight type, vee-shaped to a point at the center.
3. The vee point of the toothrest should be aligned with the vee point of the wheel and centered for height with the center gage.
4. The wheelhead drop is calculated the same way as for the plain helical mill.
5. Grind the primary clearance, first in one direction on one tooth, and then back in the opposite direction with the other tooth on the other side of the toothrest.
6. Set the drop and grind the secondary clearance, if needed. Note: Some staggered tooth cutters are made so a secondary grind is not necessary.
7. The side cutting edges are not usually ground unless they have been dulled from severe use. (Grinding the side teeth of a cutter reduces its usefulness somewhat, since a nominal, or even number, width is useful for cutting grooves of a standard width, such as keyseats.) This can be easily done, using a flaring cup wheel on a tilting wheelhead. If your machine is not so equipped, an attachment called a universal workhead, such as that shown in the textbook. a stub arbor clamped in a collet can be used. Set up the workhead to about 5 to 10 degrees clearance. If that attachment is used, a flicker type toothrest is used to position the teeth. If a cutter is narrowed by sharpening on the side, the marking on the side of the cutter that identifies its nominal width should be ground off and the new dimension put in its place with a tool marker.
8. Show your sharpened cutter to your instructor.

WORKSHEET 3. Sharpening an End Mill

Material

One or several dulled end mill

Procedure

1. Perhaps 80 to 90 percent of all end mill sharpening is done on the end since that is where most of the cutting is done. Breakage is usually on the outer corners. These corners can be removed by simply cutting off the end. See figures in the text. A damaged end may also be removed by grinding it off on a pedestal grinder, but care must be taken not to overheat the tool.
2. When a two-flute end mill has been cut off, it looks something like a thick letter "S" on the end. At this point, it must be gashed to restore the cutting edge.

3. Commercial tool sharpeners usually do this operation offhand as shown in the text. in order to save. It is a somewhat hazardous method, however, for the inexperienced tool grinder. Some school instructors prefer the students do this operation with an universal toolhead.

4. Whichever method you use, the point is to make grinds or gashes on each tooth so that they meet at the center of the cutter, leaving a slightly undercut portion at the center. You must avoid touching the outer cutting edges when you do this.

5. Set up the cutter in the universal workhead with a flicker toothrest on the notched plate to locate the tooth positions. Set the opposite cutting edges at the same height.

6. Use a flaring cup wheel, dressed to a sharp edge. Set the universal toolhead to the primary clearance angle of about 4 degrees and adjust the stop so the edge of the wheel comes to the center of the cutter.

7. Grind the primary angle on all teeth, just removing the undercut at the center. This will bring the two cutting edges to a single point at the center of the cutter.

8. Set the toolhead to the secondary angle of about 15 degrees.

9. Grind the secondary angles.

10. If the periphery of the end mill is in need of sharpening, use the procedures outlined in your textbook. Ask your instructor about this procedure.

11. Turn in the grading sheet and the cutters to your instructor for evaluation.

PROJECT 18. PIN PUNCH SET

INSTRUCTOR EVALUATION

NAME_____DATE_____

	Grade	
	Letter	Percent
1. Follows drawing (dimensions and tolerances)	_____	_____
2. Machining finishes	_____	_____
3. Mechanism or tool operates satisfactorily	_____	_____
4. General workmanship	_____	_____
Total grade	_____	_____

Comments

Instructor signature: _____

PROJECT 18. PIN PUNCH SET

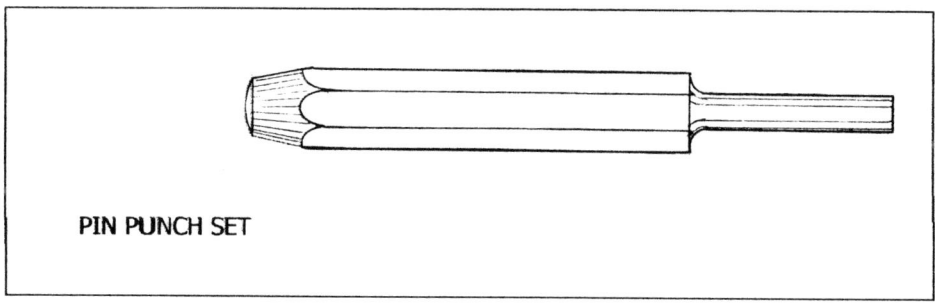

PIN PUNCH SET

Objectives

1. Learn to identify and use tool steels.
2. Learn to turn a taper using the compound rest.
3. Be able to harden and temper a tool steel correctly.

Outline for Study

Prior to starting each procedure for this project, study and complete Post-Tests for applicable units in:

1. Identifying metals: **Section D**
2. Machining the punches: **Section I**
3. Hardening and tempering the punches: **Section D**

Procedures

Worksheet 1. Machining the Punches
Worksheet 2. Hardening and Tempering the Punches

WORKSHEET 1. Machining the Punches

Materials

15 in. of 3/8 in. dia. hexagonal or octagonal S-5 tool steel or SAE 1095 carbon steel 11 in. of 1/2 in. dia. hexagonal or octagonal S-5 tool steel or SAE 1095 carbon steel

Procedure

Begin by selecting the material from the steel rack. It can be 3/8 in. and 1/2 in. octagon or hexagon stock as measured across the flats. Be sure it is tool steel and not mild steel; if you are not sure, as your instructor or spark test a sample if you are familiar with spark testing procedures. This tool steel may be S-5 shock resistant tool steel or SAE 1095 which contains sufficient carbon to harden by quenching. In the as-rolled condition, this tool steel may have hard spots and so will quickly dull a band saw. For this reason, you should cut off this material with a hand hacksaw. Cut off all eight pieces 1/16 in. over the given length on the drawing.

1. Set up a four-jaw chuck if the stock is 8-sided, or a three-jaw chuck if it is 6-sided.

2. Extend the piece out from the chuck the length of the part to be machined, plus clearance for the tool holder. Set up the stock to run true in the chuck.

3. Take fairly light cuts with a right-hand roughing tool since the work overhang prohibits deep cuts. Special care must be taken on the end of the No. 1 punch. It finishes to 0.061 in., so if you are taking even very light cuts, the last cut often breaks the end off. A method that works better is to first reduce the diameter to about 1/4 in. with a fairly sharp-nosed roughing tool. Then take several cuts for a short distance (about 1/4 in.) along the punch end, leaving the diameter about 0.001 in. over the finish size. Then proceed along the next 1/4 in., reducing the remaining length of the punch end in the same manner. File _very_ lightly and carefully to size, rotating the part at a high RPM. If you bend it while filing, it will break off. Be sure to leave enough material for the radius.

4. Grind a 1/8 in. radius tool. Check it with a radius gage. This tool should have zero rake and a standard relief angle. It should be set up on center. A fillet radius is one of the most important cutting operations that a machinist must be able to do correctly. It should not be an undercut as this should form a groove that could weaken the part and cause it to break in service. It should be smooth with no tool marks which can cause cracks to develop in this sensitive location.

5. Using a lower speed than for single point turning and cutting oil, cut the radius. Machine the shoulder to length (if you are working on the No. 4 through No. 8 punches).

6. Set the compound to 10 degrees and rough out and finish the tapered part to length (if you are working on the No. 1 through No. 3 punches).

7. Turn the part end for end in the chuck and set up for running true.

8. Turn the 10 degree taper as shown in the drawing.

9. Free-hand turn the 3/8 in. radius and check with a radius gage. Finish with a file and abrasive cloth.

10. Remove the punch from the lathe and stamp your name or initials on the shank, if desired.

WORKSHEET 2. Hardening and Tempering the Punches

1. Grasp the shank end of the punches with tongs and place them on a brick or raised block in a furnace that has been heated to 1550°F.

2. When the punches have become the same color as the furnace bricks, remove one punch at a time and quench in water or oil, depending the type of tool steel. The No. 1 punch should be quenched in oil in any case since a water quench is too severe for this small mass and could cause it to crack.

3. Remove any oil from the punches and polish off the scale on the machined surfaces with abrasive cloth until a bright metallic surface shows. Be careful to keep any oil or fingerprints off this surface.

4. Place the shank part of the punches on a hot plate and, if necessary, move them around to obtain an even coloring. The shank end where it will be struck with a hammer should become violet or blue, but the punch end should color a deep gold with a mottling of purple where the punch end joins the shank.

5. When these colors are obtained, quickly plunge the punch in water so it will not become any softer.

6. Turn in the punch set for grading. Leave the tempering colors on the punches so your instructor can evaluate the heat treatment.

PROJECT 19. CNC PROGRAMMING
INSTRUCTOR EVALUATION

NAME_____DATE_____

CNC Programming Checklist   **Reviewed/Approved**

1. Machining operations listed   _____

2. Cutting tools selected and listed   _____

3. Sequence of machining operations detailed.   _____

4. Program developed.   _____

5. Tooling installed.   _____

6. Z axis tool offsets set and double checked.   _____

7. Program loaded   _____

8. Program proved out   _____

9. Workpiece properly secured.   _____

10. CNC safety rules reviewed   _____

11. Workpiece machined.   _____

12. Dimensions inspected   _____

Grade

	Letter	Percent
1. Follows drawing (dimensions and tolerances)	_____	_____
2. Machining finishes	_____	_____
3. Part meets tolerances	_____	_____
4. General workmanship	_____	_____
Total grade	_____	_____

Comments

Instructor signature _____

PROJECT 19. CNC PROGRAMMING
Linear and Circular Interpolation Milling, Drilling and Tapping

Objectives

1. Demonstrate your ability to plan, develop and prove out a basic CNC program for the machining center involving linear and circular interpolation milling, point to point drilling and tapping, using industry standard conventional programming code (general Fanuc or compatible CNC pattern).

2. Select appropriate cutting tools for the job.

3. Set up the machining center tooling and establish Z axis tool offsets.

4. Prove out program on computer or machine using the Z axis inhibit feature.

5. Machine part and inspect dimensions for accuracy

Outline for Study

1. Study appropriate units in **Section M** pertaining to CNC dimensioning systems and programming code.

2. Complete the following worksheet. (Refer to drawing **CNC-1**)

3. Review any unique or specific CNC coding for the equipment that you are using.

Worksheet

1. Select, list and sequence all the cutting tools required for this project.

   Tool #   **Tool name**

   1. _____

   2. _____

   3. _____

   4. _____

   5. _____

2. Develop the CNC program for the project. Include any required insurance blocks and program reset to the beginning when machining sequences are complete.

N10 _____

N20 _____

N30 _____

N40 _____

N50 _____

N60 _____

N70 _____

N80 _____

N90 _____

N100 _____

N110 _____

N120 _____

N130 _____

N140 _____

N150 _____

N160 _____

N170 _____

N180 _____

N190 _____

N200 _____

(Use additional programming blocks as needed).

3. Install cutting tools in the tool changer and set up Z axis offsets for each tool.

4. Have you instructor **check and verify** Z axis tool offset settings on the machining center.

5. Load and prove out you program on CNC machine or computer and verify.

6. Secure workpiece and machine part.

7. Check for correct dimensions.

8. Submit part and evaluation sheet to you instructor.

CNC-1 PROGRAMMING PROJECT

MATL 6061 T6 ALUM TOOLING PLATE

PROJECT 20. CNC PROGRAMMING
INSTRUCTOR EVALUATION

NAME_____DATE_____

CNC Programming Checklist   **Reviewed/Approved**

1. Machining operations listed   _____

2. Cutting tool selected and listed   _____

3. Sequence of machining operations detailed.   _____

4. Program developed.   _____

5. Tooling installed.   _____

6. Z axis tool offsets set and double checked.   _____

7. Program loaded   _____

8. Program proved out   _____

9. Workpiece properly secured.   _____

10. CNC safety rules reviewed   _____

11. Workpiece machined.   _____

12. Dimensions inspected   _____

Grade

	Letter	Percent
1. Follows drawing (dimensions and tolerances)	_____	_____
2. Machining finishes	_____	_____
3. Part meets tolerances	_____	_____
4. General workmanship	_____	_____
Total grade	_____	_____

Comments

Instructor signature _____

PROJECT 20. CNC PROGRAMMING
Coordinate Drilling

Objectives

1. Demonstrate your ability to plan, develop and prove out a basic CNC program for the machining center involving coordinate drilling using industry standard conventional programming code (general Fanuc or compatible CNC pattern).

2. Select appropriate cutting tools for the job.

3. Set up the machining center tooling and establish Z axis tool offsets.

4. Prove out program on computer or machine using the Z axis inhibit feature.

5. Machine part and inspect dimensions for accuracy

Outline for Study

1. Study appropriate units in **Section M** pertaining to CNC dimensioning systems and programming code.

2. Complete the following worksheet. (Refer to drawing **CNC-2**)

3. Review any unique or specific CNC coding for the equipment that you are using.

Worksheet

1. Select, list and sequence all the cutting tools required for this project.

Tool #	Tool name
1.	_____
2.	_____
3.	_____
4.	_____

2. Using the center of the pitch circle as absolute zero, calculate the coordinates for the drilled features and enter these dimensions in the coordinate table below. Refer to drawing **CNC-2**

Hole #	X Axis coordinate	Y Axis coordinate
1		
2		
3		
4		
5		

3. Develop the CNC program for the project. Include any required insurance blocks and program reset to the beginning when machining sequences are complete.

N10 _____

N20 _____

N30 _____

N40 _____

N50 _____

N60 _____

N70 _____

N80 _____

N90 _____

N100 _____

N110 _____

N120 _____

N130 _____

N140 _____

N150 _____

N160 _____

N170 _____

N180 _____

N190 _____

N200 _____

(Use additional programming blocks as needed).

    3. Install cutting tools in the tool changer and set up Z axis offsets for each tool.

    4. Have you instructor **check and verify** Z axis tool offset settings on the machining center.

    5. Load and prove out you program on CNC machine or computer and verify.

    6. Secure workpiece and machine part.

    7. Check for correct dimensions.

    8. Submit part and evaluation sheet to you instructor.

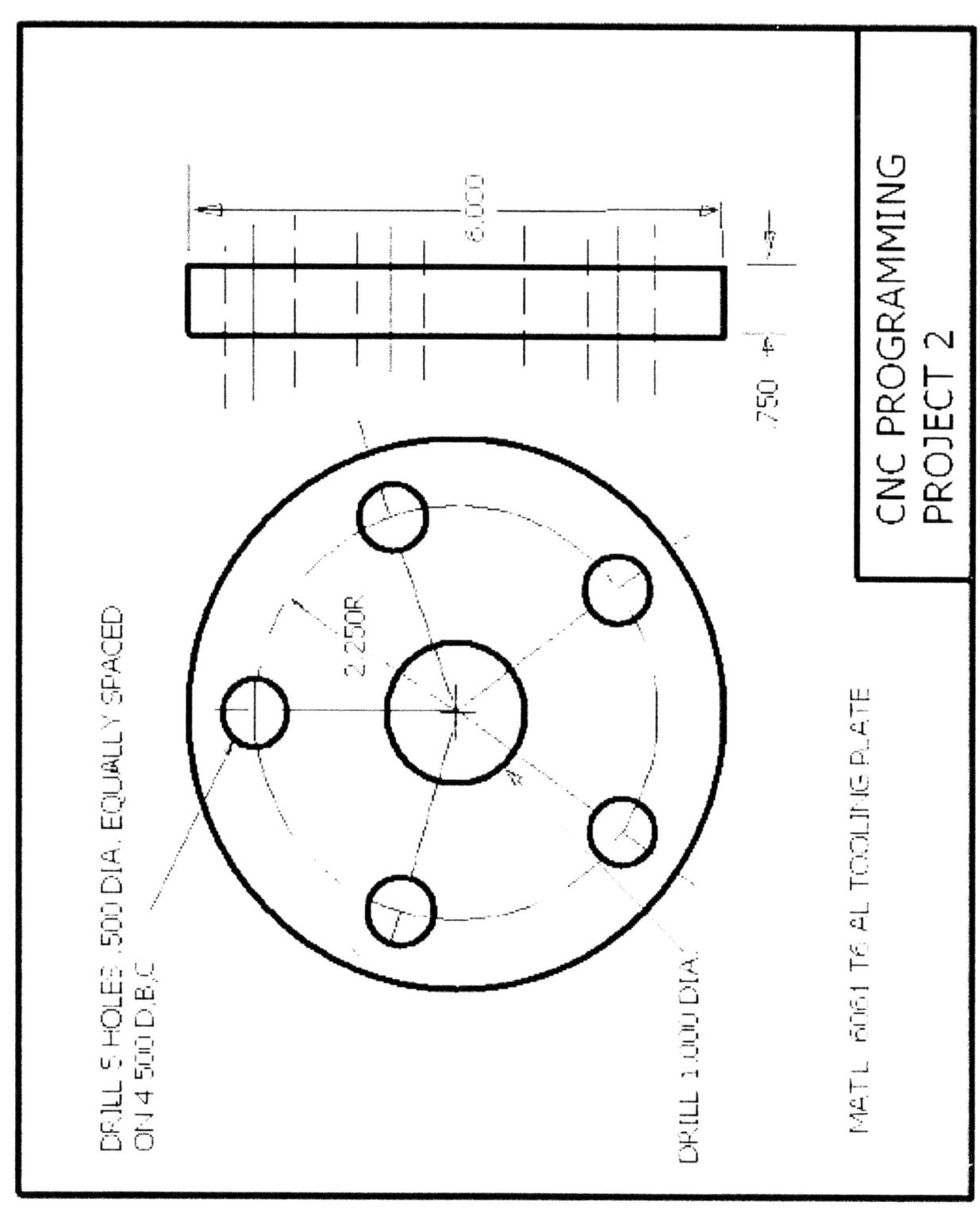

ADDITIONAL PROJECTS

1. WHEEL PULLER

2. MACHINE VISE

ADDITIONAL PROJECT 1. WHEEL PULLER

INSTRUCTOR EVALUATION

NAME_____DATE_____

Grade

	Letter	Percent
1. Follows drawing (dimensions and tolerances)	_____	_____
2. Machining finishes	_____	_____
3. Mechanism or tool operates satisfactorily	_____	_____
4. General workmanship	_____	_____
Total grade	_____	_____

Comments

Instructor signature: _____

PROJECT A-1. WHEEL PULLER

WHEEL PULLER

Objectives

1. Be able to turn a disc shape on a lathe and tap an internal thread.
2. Learn to mill slots on a horizontal milling machine.
3. Be able to use a vertical bandsaw.
4. Use a steady rest on a lathe.
5. Case harden mild steel.

Outline for Study

Prior to starting each procedure for this project, study and complete Post-Tests for applicable units in:

1. Making the body: **Section J, Section K**
2. Making the nut: **Section I**
3. Making the screw: **Section I**
4. Making the legs: **Section G and Section D**

Procedures

Begin the procedures for this project by completing the following worksheets.

Worksheet 1. Making the Body

Worksheet 2. Making the Nut

Worksheet 3. Making the Screw

Worksheet 4. Making the Legs

WORKSHEET 1. Making the Body

Materials

Three 1/4-20 UNC cap screws   7/8 in. 3-1/2 in. 3-1/2 in. HRMS round bar

168 **Procedure**

1. Saw the material for the body from 3-1/2 in. MS round stock. Cut off a piece about 7/8 in. long.

2. Set up the three-jaw chuck with the jaws set to grip the outside of this piece. Place the material in the chuck with the jaws gripping about 1/8 in. True up the piece by measuring from the chuck face with a rule in several locations.

3. Take a facing cut to clean up one side.

4. True the OD to the size specified on the drawing. Machine as close to the chuck jaws as possible, staying clear of the jaws with the tool. Chamfer the edge.

5. Remove the part from the chuck and apply layout dye to the finished surface.

6. To layout the part, find the center and scribe a centerline "a" across the surface (Figure 1). Prick punch the center.

7. Set the dividers to approximately 1-1/2 in. Out one leg in the prick punch mark and scribe san arc of about 180 degrees.

8. With the dividers at the same setting, place one leg where the centerline and the arc are coincident. this will mark points 60 degrees on the other side of the centerline ("d" and "e" in Figure 2).

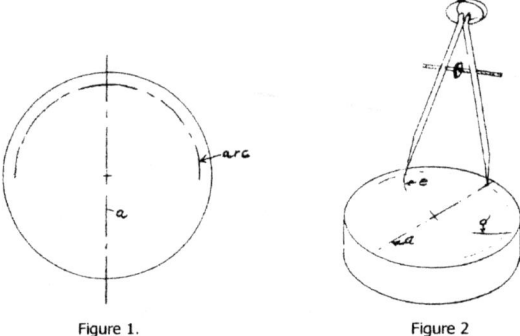

Figure 1.   Figure 2

9. Scribe a centerline ("b" and "c", Figure 3) through the center to the two short arcs ("d" and "e").

10. On the four lines that now extend to the outside edge of the circle, measure inward 7/8 in. and scribe a short line perpendicular to the centerline to outline the bottom of the slots.

11. Measure 1/8 in. on either side of the centerline and scribe parallel liners to meet the short cross line (Figure 3).

12. Scribe parallel lines 3-1/2 in. from either side of the slots. Mark their length from the outer edge 11/16 in.

13. Connect the ends of these lines to form the remainder of the outline of this part. The layout is now complete.

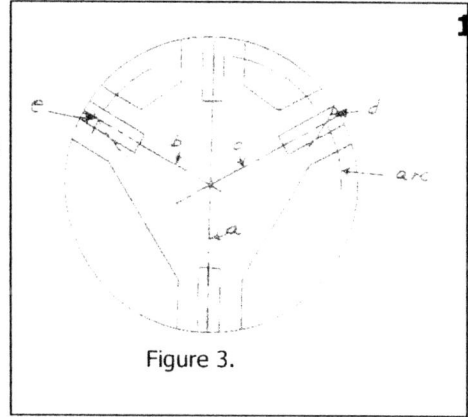

Figure 3.

14. Replace the part in the lathe chuck with the finished face in. Grasp it this time as far back into the jaws as it will go. Be sure the seating surfaces of the jaws are clean and free from chips.

15. Face to the thickness given in the drawing. This should remove the 1/8 in. section where the jaws were holding the part in the last operation. Chamfer the OD.

16. Center drill and then drill through with a pilot drill or a drill slightly smaller than the tap drill.

17. Drill through with a tap drill for 3/4-16 UNF thread.

18. Chamfer the hole.

19. Set up the tap with the dead center supporting it and the tap handle resting on the carriage or on the compound, not on the ways. Release the spindle so it is free wheeling. Turn the chuck by hand to make the thread while following with the dead center by gently turning the tailstock hand wheel. <u>Do not turn on the machine for this operation.</u> Use sulfurized cutting oil. Remove the work when finished and clean the machine.

20. To continue with this project, you should be familiar with the use of a small horizontal milling machine. You will need to be able to set up the cutter, vise, and work. You must also set your speeds and feeds. You will need to study applicable units in Section J

21. Cut a slot first, then make the two cuts on each side of the slot (Figure 4). Use cutting fluid. Some machines are capable of doing this in one cut; however, some light duty machines *may* only be capable of .200 to .400 in. depth per pass.

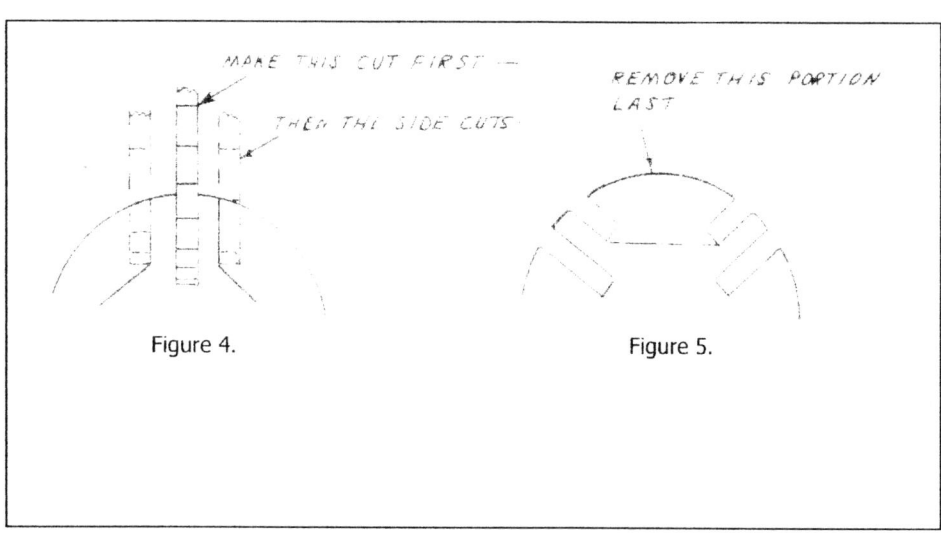

Figure 4.   Figure 5.

22. Turn the work and set up the next centerline with a square head. Repeat the cutting procedure in step 21.
23. When all four slots and side cuts have been made, rotate the work so that the line on the spaces in between is level or parallel with the solid jaw of the vise.
24. Remove all excess material by taking cuts with the same milling cutter (Figure 5).
    Note: An alternate method is to saw out the excess material on a vertical handsaw.
25. When the milling is finished, remove the part, deburr and chamfer all edges.
26. Lay out for the four holes in the ears for the cap screws and center punch them.
27. Set up in the drill press with one ear level with the table.
28. Drill through both halves of the ear with a 1/4-20 UNC tap drill (#7).
29. Drill through the top ear only with a 1/4 in. drill.
30. Rotate the piece to the next ear and repeat steps 27 through 29.
31. After all four ears are drilled, deburr and tap through the ears with a 1/4-20 UNC tap. Clean and apply a light film of oil.

WORKSHEET 2. Making the Nut

Material

1-1/4 inches of 2-1/2 HRMS round

Procedure

1. Cut off a piece of mild steel round stock slightly larger than 2-1/4 in. diameter.

   The next size larger may be 22 in. HRMS round bar. The length should be 1-1/4 in.

2. Since there are some very heavy tool pressures in this operation such as knurling, grooving, and drilling, it is best to set up the work in a four-jaw chuck. Leave 7/8 in. extending out of the chuck and roughly true up the work with no more than 0.010 in. runout.
3. The operation with the greatest work force should be done first so that if the work moves it can be more easily corrected. Therefore, the knurling should be done first, but the part must first be machined to the knurl diameter.
4. When you have turned the nut to the knurl diameter in as far as the shoulder (9/16 in.),

   then set up a medium knurl and knurl the part.
5. Grind a tool bit with a 1/8 in. radius.
6. Machine the groove with the lowest speed on the lathe. Use cutting oil. This plunge cut, like parting off, may produce some chatter. If the feed is too light, it will almost always chatter, so feed in by hand at a rate that will just produce a chip. Too much feed will also cause chatter or jamming of the tool in the work, causing the work to come out of the chuck or the machine to stop.

7. Face the piece and chamfer both sides of the knurled part. Finish turn both faces and the shoulder again to 9/16 in. from the end to correct any runout.
8. Center drill, pilot drill, and tap drill with the same procedure as with the body.
9. Tap 3/4-16 UNF.
10. Remove the nut from the chuck and change to a three-jaw chuck.
11. Chuck a threaded stub mandrel having a 3/4-16 UNF thread.
12. Screw on the nut with the knurled part next to the chuck.

13. Turn the OD to 2-1/4 in. Face to the 13/16 in. length. Mark a 2 in. diameter line on the face
14. Free hand machine a slight radius from the OD to the 2 in. line as shown on the drawing.
15. Finish the radius with a file and deburr the edge nearest the chuck. Chamfer the threaded hole. Clean and apply a light film of oil.

WORKSHEET 3. Making the Screw

Material

7-1/2 inches of 1 in. hexagonal stock

Procedure

1. Cut off a piece of 1 in. hexagonal stock 7-1/2 in. long. The extra length is for turning off the center.
2. Chuck the piece in a three-jaw chuck and face both ends. Chamfer one end to a point slightly below the flats on the hexagon stock. Center drill the other end. Chuck 1/2 in. of the chamfered end and support the other end in the tailstock center.
3. Turn the diameter to size. Check for tapering and correct if needed.
4. Set up for threading and cut the 3/4-16 UNF thread. Check for fit with the nut you have just made. Turn the 1/2 in. diameter. This should now be about 3/8 in. longer than it will be when finished.
5. With one end still in the chuck, set up a steady rest on the other end at the 1/2 in. diameter. Move the tailstock away. Be sure the carriage is to the right of the steady rest so you can turn the end of the piece.
6. Machine off the center hole and face to length, being careful to leave material in the center for the point.
7. Remove the part from the lathe and case harden by the roll method explained in Section D.
8. Clean up the part and apply light oil.

WORKSHEET 4. Making the Legs

Material

18-1/2 inches of 1/4 x 1 in. HRMS

Procedure

1. Cut off three pieces of 1/4 x 1 in. MS 6-1/16 in. long.
2. Lay out as shown on the drawing.
3. Drill the 1/4 in. holes. Chamfer.
4. Saw along the layout lines with the vertical band saw, leaving a minimum of material to file. Use a push stick for safety when sawing.
5. Finish file the legs and chamfer all edges.
6. Case harden all three legs by the roll method. Clean and apply a light film of oil.

Procedure for Assembly

1. Obtain three 1/4-20 UNC socket head cap screws 3/4 in. long. Assemble the legs on the body.

2. Install the nut on the screw and turn the screw into the body.

3. This tool may also be used in the alternate position with two legs opposed.

4. Turn in your grading sheet and completed wheel puller for your instructor's evaluation.

ADDITIONAL PROJECT 2. MACHINE VISE

INSTRUCTOR EVALUATION

NAME_____DATE_____

Grade

	Letter	Percent
1. Follows drawing (dimensions and tolerances)	_____	_____
2. Machining finishes	_____	_____
3. Mechanism or tool operates satisfactorily	_____	_____
4. General workmanship	_____	_____
Total grade	_____	_____

Comments

Instructor signature: _____

PROJECT A-2 MACHINE VISE

Objectives

1. Learn to square a block and to machine steps on a horizontal mill.
2. Be able to drill and ream holes and machine slots on a vertical mill.
3. Surface grind the jaw inserts.
4. Be able to make an Acme screw and nut.

Outline for Study

Prior to starting each procedure for this project, study and complete Post-Tests for applicable units in:

1. Machining the vise body: **Section K**
2. Milling the slots and pocket: **Section J**
3. Making the movable jaw: **Section J**.
4. Grinding the jaw inserts: **Section L**
5. Making the screw and nut: **Section I**

Procedures

Begin the procedures for this project by completing the following worksheets.

Worksheet   1. Machining the Vise Body
Worksheet   2. Milling the Slots and Pocket
Worksheet   3. Making the Movable Jaw and Inserts
Worksheet   4. Grinding the Jaw Inserts
Worksheet   5. Making the Screw and Nut

WORKSHEET 1. Machining the Vise Body

Materials

8-1/4 inches of 2 x 4 in. HRMS bar   8 inches of 1/2 x 1-1/2 in. HRMS bar

Procedure

1. Cut off the 2 by 4 inch stock and two pieces of $_2$ by 12 inch bar 4 inches **long.**

   (Drawings for this project may be found at the end of these worksheets.)
2. Set up the block in a horizontal mill and square it as explained in Section K. Leave it oversize for finishing, about 0.050 in.
3. Square the ends, leaving the block about 0.050 in. long for finishing cuts.
4. Make the chamfer for the welds on both ends of the block.
5. Mill the chamfer on both hold-down straps and machine the ends square and to the oversize width of the block. Do not cut the slots in the straps at this time.

6. Clamp the straps and block to a flat surface and tack weld the upper side. The block should be preheated before welding on it because of its mass. Turn it over and weld in the groove. Alternate passes on each side to prevent warpage.
7. Set the block back in the milling vise for milling out the center portion. This is best done with a large staggered tooth cutter, about 3/4 in. wide.
8. Two or more roughing cuts may be needed to machine to depth. Leave 0.030 in for finish. Make the finish cuts. If possible in the same setup, take a clean up cut on the top of the hold-down straps, leaving the weld intact. Also take a side cut to finish the outside end of the vise body, _leaving the weld intact. Hold all dimensions within a tolerance of plus or minus 0.005 in.
9. Turn the vise body over and support the surface you have just machined with parallels.
10. Make the finish cut on the bottom, cleaning up the weld.
11. Set up and finish mill the sides.
12. The vise body may now be taken to a radial arm drill press and clamped to an angle plate or to the side of the drill table. The 7/8 hole in one end may now be drilled and reamed. Be sure to chamfer both ends of the hole.
13. Drill and counterbore the other end. The vise surface is completed. No further finishing (such as surface grinding) should be needed, except for deburring and chamfering the sharp edges. The holes for the jaw insert bolts may be drilled together with the inserts to insure alignment.

WORKSHEET 2. Milling the Slots and Pocket

Tools and Materials

Vertical milling machine   3/4 in. center-cutting end mill

Procedure

1. The vise body should now be taken to a vertical milling machine and set up in a milling vise, bottom up.
2. The slots in the hold-down clamps may now be cut and also the 3/ in. slot in the vise body. The pocket should also be milled in this setting.
3. Turn the vise body on its side and mill the hold-down slots.
4. This completes the vise body.

WORKSHEET 3. Making the Movable Jaw and Inserts

Materials

4 inches of 1-1/2 x 2 in. HRMS bar   8 inches of 5/16 x 1-1/2 in. spring steel flat bar or equivalent

Procedure for Movable Jaw

1. Square the bar on the horizontal mill as shown in your text on pages 579 to 582. Leave it slightly oversize.
2. Finish the bottom side and mill the steps. Make sure the inch guide fits the slot you have milled in the body. Measure the slot and make the guide 0.003 in. smaller for a sliding fit.
3. Finish milling the other sides and ends to the dimensions given on the drawing. Hold to a tolerance of ± 0.003 in.

4. The drilling, tapping, and countersinking may be done on the vertical mill or on a drill press. The 9/16 in. socket for the screw should be reamed to size. If you wish, you may wait to drill the holes that hold the jaw inserts until the inserts are finished; they can be clamped together and drilled for better alignment.

   Drill the 1/4 in. pin holes and their 1/8 in. knockout holes before the 9/16 in. hole is drilled.
5. The movable jaw is now finished.

Procedure for the Jaw Inserts

1. Since the material for the jaw inserts is spring steel, it tends to workharden very quickly if the cutting speed is too high, the feed is too low, or the tool is dull. The cutting speed for this material should be about 30 to 40 FPM ; On many machines, this is the lowest possible RPM. Use cutting oil.
2. Cut 8 inches of bar and leave it this length until all the milling is done on the sides.
3. Set up on a horizontal mill in a vise for milling the grooves.
4. Take a clean-up cut with a slab mill; then change to a 60 degree single angle cutter or a 120 degree double angle cutter if one is available.
5. With a single angle cutter, one side of the angle is first cut, then the cutter is reversed on the arbor and the other side is cut.
6. The vise can be rotated 90 degrees and the short vee grooves can be cut without removing the work from the vise.
7. Now change to a side cutting cutter, and set up the bar on its edge. Mill the sides.
8. Now mill the step, measuring from the finished side.
9. Set up to mill the other flat side that is still unfinished and mill it to size, leaving about 0.020 in. for grinding.
10. Saw the bar into two parts and mill the ends square and to length.
11. Lay out, drill, and tap the 1/4-28 UNF holes. The drilling of the movable jaw and vise body may be done together.
12. All of the machining must be done before the next step.
13. Harden and temper the jaws as explained in Section D of your textbook.

Procedure for Retainer

1. At this point the retainer plate should be made as shown on the drawing. It should slide freely in the pocket in the vise body and not extend below the bottom of the vise body.
2. The 1/4 inch holes should be drilled and countersunk so that the screw heads are sunk below the surface.

WORKSHEET 4. Grinding the Jaw Inserts

Materials

Surface grinding machine and accessories

Procedure

1. Grind the flat sides of the jaw inserts as explained in Section L of the textbook.

2. Clamp the inserts into a precision vise or an angle plate and grind the edges square and to size.

WORKSHEET 5. Making the Screw and Nut

Materials

8-1/8 inches of 3/4 in. dia. CR round   1 inch of 1-1/4 in. round bronze stock 4 inches 4 inches of f 3/4 x 3/4 in. keystock (for optional lever)

Procedure

1. Centerdrill both ends of the 3/4 in. round bar.
2. Set up in a dividing head for milling the four flats. If the optional lever is to be made, then simply mill two flats as shown on the drawing and drill the hole.
3. Set up in a lathe with a four-jaw chuck on the square end.
4. Turn down the $_{16}$ inch end and make the groove.
5. Make the Acme thread as explained in **Section I**
6. Set up a short piece of bronze bar stock in the lathe and turn the OD to be a light press fit in the vise body.
7. Tap drill for a 3/4-6 Acme thread.
8. Part off the nut and set it up again in a three-jaw chuck.
9. Run a set of Acme taps through the nut while it is chucked in the lathe. Do not use power, turn the chuck or tap by hand. Alternately, if there is no tap, make the thread with a single point Acme tool.

Assembly

1. Clean and oil the parts.
2. Press the nut into the vise body.
3. Bolt the insert jaws to their parts. Fasten the movable jaw and screw.
4. Take the grading sheet and finished vise to your instructor for evaluation.

Table 1

Decimal and Metric Equivalents of Fractions of an Inch

Fractional Inch	Decimal Inch	Millimeters	Fractional Inch	Decimal Inch	Millimeters
1/64	0.015625	0.3969	33/64	0.515625	13.0969
1/32	0.03125	0.7937	17/32	0.53125	13.4937
3/64	0.046875	1.1906	35/64	0.546875	13.8906
1/16	0.0625	1.5875	9/16	0.5625	14.2875
5/64	0.078125	1.9844	37/64	0.578125	14.6844
3/32	0.09375	2.3812	19/32	0.59375	15.0812
7/64	0.109375	2.7781	39/64	0.609375	15.4781
1/8	0.125	3.1750	5/8	0.625	15.8750
9/64	0.140625	3.5719	41/64	0.640625	16.2719
5/32	0.15625	3.9687	21/32	0.65625	16.6687
11/64	0.171875	4.3656	43/64	0.671875	17.0656
3/16	0.1875	4.7625	11/16	0.6875	17.4625
13/64	0.203125	5.1594	45/64	0.703125	17.8594
7/32	0.21875	5.5562	23/32	0.71875	18.2562
15/64	0.234375	5.9531	47/64	0.734375	18.6531
1/4	0.25	6.3500	3/4	0.75	19.0500
17/64	0.265625	6.7469	49/64	0.765625	19.4469
9/32	0.28125	7.1437	25/32	0.78125	19.8437
19/64	0.296875	7.5406	51/64	0.796875	20.2406
5/16	0.3125	7.9375	13/16	0.8125	20.6375
21/64	0.328125	8.3344	53/64	0.828125	21.0344
11/32	0.34375	8.7312	27/32	0.84375	21.4312
23/64	0.359375	9.1281	55/64	0.859375	21.8281
3/8	0.375	9.5250	7/8	0.875	22.2250
25/64	0.390625	9.9219	57/64	0.890625	22.6219
13/32	0.40625	10.3187	29/32	0.90625	23.0187
27/64	0.421875	10.7156	59/64	0.921875	23.4156
7/16	0.4375	11.1125	15/16	0.9375	23.8125
29/64	0.453125	11.5094	61/64	0.953125	24.2094
15/32	0.46875	11.9062	31/32	0.96875	24.6062
31/64	0.484375	12.3031	63/64	0.984375	25.0031
1/2	0.50	12.7000	1	1.000000	25.4000

Table 2

Tap Drill Sizes for Unified and American Standard Series Screw Threads

Thread Size	Threads Per Inch	Series	Tap Drill Diameter Size	Inches	Percent of Full Thread	Thread Size	Threads Per Inch	Series	Tap Drill Diameter Size	Inches	Percent of Full Thread
0	80	NF	1.25 mm	.0492	66	6	40	NF	31	.1200	55
			1.2 mm	.0472	79				32	.1160	68
			3/64"	.0469	81				33	.1130	77
1	64	NC	1/16"	.0625	51				34	.1110	83
			53	.0595	66	8	32	NC	28	.1405	58
			54	.0550	88				29	.1360	69
1	72	NF	1/16"	.0625	58				3.4 mm	.1338	74
			1.55 mm	.0610	66				3.3 mm	.1299	84
			53	.0595	75	8	36	NF	27	.1440	55
2	56	NC	49	.0730	56				28	.1405	65
			50	.0700	69				29	.1360	77
			1.75 mm	.0689	74				3.4 mm	.1338	83
			51	.0670	82	10	24	NC	22	.1570	61
2	64	NF	1.9 mm	.0748	55				24	.1520	70
			49	.0730	64				25	.1495	75
			1.8 mm	.0709	74				26	.1470	79
			50	.0700	79	10	32	NF	19	.1660	59
3	48	NC	45	.0820	63				20	.1610	71
			46	.0810	66				21	.1590	76
			47	.0785	76				22	.1570	81
			48	.0760	85	12	28	NF	14	.1820	63
3	56	NF	44	.0860	56				15	.1800	66
			2.15 mm	.0846	62				16	.1770	72
			45	.0820	73				17	.1730	79
			46	.0810	77	12	28	NF	12	.1890	58
4	40	NC	42	.0935	57				14	.1820	73
			2.3 mm	.0905	66				15	.1800	77
			43	.0890	71				16	.1770	84
			44	.0860	80	12	32	NEF	3/16"	.1875	70
4	48	NF	41	.0960	59				13	.1850	76
			42	.0935	68				14	.1820	84
			2.3 mm	.0905	79	1/4"	20	UNC	5	.2055	67
			43	.0890	85			NC	6	.2040	71
5	40	NC	37	.1040	65				7	.2010	75
			38	.1015	72				8	.1990	78
			39	.0995	78	1/4"	28	UNF	2	.2210	62
			40	.0980	83			NF	7/32"	.2187	67
5	44	NF	36	.1065	62				5.5 mm	.2165	72
			37	.1040	71				3	.2130	80
			38	.1015	79	1/4"	32	NEF	5.7 mm	.2244	63
			39	.0995	86				2	.2210	71
6	32	NC	34	.1110	66				7/32"	.2187	77
			35	.1100	69				5.5 mm	.2165	82
			36	.1065	77	5/16"	18	UNC	17/64"	.2656	65
			37	.1040	83			NC	G	.2610	71

Table 2 (continued)

Thread Size	Threads Per Inch	Series	Tap Drill Diameter Size	Inches	Percent of Full Thread
			F	.2570	77
			6.4 mm	.2520	84
5/16"	24	UNF NF	J	.2770	66
			I	.2720	75
			H	.2660	85
5/16"	32	NEF	7.3 mm	.2874	62
			7.2 mm	.2835	71
			9/32"	.2812	77
			J	.2770	87
3/8"	16	UNC NC	P	.3230	64
			O	.3160	72
			5/16"	.3120	77
			7.8 mm	.3071	83
3/8"	24	UNF NF	R	.3390	67
			8.5 mm	.3346	74
			Q	.3320	79
			21/64"	.3281	86
7/16"	14	UNC NC	3/8"	.3750	67
			U	.3680	75
			23/64"	.3594	84
7/16"	20	UNF NF	X	.3970	62
			25/64"	.3906	72
			W	.3860	79
1/2"	13	UNC NC	7/16"	.4375	62
			27/64"	.4219	78
			Z	.4130	87
1/2"	20	UNF NF	11.75 mm	.4626	57
			29/64"	.4531	72
9/16"	12	UNC NC	1/2"	.5000	58
			31/64"	.4844	72
			15/32"	.4687	86
9/16"	18	UNF NF	33/64"	.5156	65
			13 mm	.5118	70
			1/2"	.5000	86
5/8"	11	UNC NC	14 mm	.5512	63
			35/64"	.5469	66
			17/32"	.5312	79
5/8"	18	UNF NF	37/64"	.5781	65
			14.5 mm	.5709	75
			9/16"	.5625	87
3/4"	10	UNC NC	17 mm	.6693	62
			21/32"	.6562	72
			41/64"	.6406	84
3/4"	16	UNF NF	45/64"	.7031	58
			11/16"	.6875	77
			43/64"	.6719	96
7/8"	9	UNC NC	25/32"	.7812	65
			49/64"	.7656	76
			3/4"	.7500	87
7/8"	14	UNF NF	13/16"	.8125	67
			20.5 mm	.8071	73
			51/64"	.7969	84
1"	8	UNC NC	57/64"	.8906	67
			7/8"	.8750	77
			55/64"	.8593	87
1"	12	UNF N	15/16"	.9375	58
			59/64"	.9218	72
			29/32"	.9062	87

Some Symbols Used for American Threads Are:

Symbol	Reference
NC	American National Coarse Thread Series
NF	American National Fine Thread Series
NEF	American National Extra Fine Thread Series
NS	Special Threads of American National Form
NH	Am. Natl. Hose Coupling and Fire Hose Coupling Thread
NPT	American Standard Taper Pipe Thread
NPTF	American Standard Taper Pipe Thread (Dryseal)
NPS	American Standard Straight Pipe Thread
ACME	Acme Threads—(Acme-C) Centralizing—(Acme-G) General Purpose
STUB ACME	Stub Acme Threads
V	A 60° "V" thread with truncated crests and roots. The theoretical "V" form is usually flattened several thousandths of an inch
SB	Manufacturers Stovebolt Standard Thread

Symbols Used for Unified Threads Are:

Symbol	Reference
UNC	Unified Coarse Thread Series
UNF	Unified Fine Thread Series
UNEF	Unified Extra Fine Thread Series

Table 3

Metric-Inch Conversion Table

Millimeters	Decimal Inches	Millimeters	Decimal Inches	Millimeters	Decimal Inches	Millimeters	Decimal Inches	Millimeters	Decimal Inches	Millimeters	Decimal Inches
.1	.00394	11	.4331	29	1.1417	47	1.8504	65	2.5590	83	3.2677
.2	.00787	12	.4724	30	1.1811	48	1.8898	66	2.5984	84	3.3071
.3	.01181	13	.5118	31	1.2205	49	1.9291	67	2.6378	85	3.3464
.4	.01575	14	.5512	32	1.2598	50	1.9685	68	2.6772	86	3.3858
.5	.01968	15	.5905	33	1.2992	51	2.0079	69	2.7165	87	3.4252
.6	.02362	16	.6299	34	1.3386	52	2.0472	70	2.7559	88	3.4646
.7	.0275	17	.6693	35	1.3779	53	2.0866	71	2.7953	89	3.5039
.8	.0315	18	.7087	36	1.4173	54	2.1260	72	2.8346	90	3.5433
.9	.03541	19	.7480	37	1.4567	55	2.1653	73	2.8740	91	3.5827
1	.0394	20	.7874	38	1.4961	56	2.2047	74	2.9134	92	3.6220
2	.0787	21	.8268	39	1.5354	57	2.2441	75	2.9527	93	3.6614
3	.1181	22	.8661	40	1.5748	58	2.2835	76	2.9921	94	3.7008
4	.1575	23	.9055	41	1.6142	59	2.3228	77	3.0315	95	3.7401
5	.1968	24	.9449	42	1.6535	60	2.3622	78	3.0709	96	3.7795
6	.2362	25	.9842	43	1.6929	61	2.4016	79	3.1102	97	3.8189
7	.2756	26	1.0236	44	1.7323	62	2.4409	80	3.1496	98	3.8583
8	.3150	27	1.0630	45	1.7716	63	2.4803	81	3.1890	99	3.8976
9	.3543	28	1.1024	46	1.8110	64	2.5197	82	3.2283	100	3.9370
10	.3937										

Table 4

Inch-Metric Measures

Linear Measures

English
1 mile = 1760 yards = 5280 feet
1 yard = 3 feet = 36 inches
1 foot = 12 inches
1 inch = 1000 mils

Metric
10 millimeters (mm) = 1 centimeter (cm)
10 centimeters = 1 decimeter (dm)
10 decimeters = 1 meter (m)
1000 meters = 1 kilometer (km)

Conversion Factors
1 inch = 2.54 cm = 25.4 mm
1 foot = .3048 meter
1 yard = .9144 meter

1 millimeter = .03937 inch
10 millimeters (cm) = .3937 inch
1 meter = 39.37 inches
           3.2808 feet
           1.0936 yards

1 mile = 1.6093 km

1 kilometer = 1093.6 yards or .62137 mile

Table 5

Useful Formulas for Finding Areas and Dimensions of Geometric Figures

A = Area; S = Side; D = Diagonal of circumscribed circle; d = Height of circular segment (distance cut into round stock to make 3 flats); h = Height of triangle.

Triangle (equilateral)

$A = Sh/2$
$S = D \times .866$
$d = D \times .25$
$h = D - d$

A = Area; S = Side; D = Diagonal; d = Height of circular segment (distance cut into round stock to make 4 flats).

Square

$A = S^2 = \frac{1}{2}d^2$
$S = D \times .7071$
$D = S \times 1.4142$
$d = D \times .14645$

A = Area; S = Side; D = Diagonal; R = Radius of circumscribed circle; r = Radius of inscribed circle; d = Height of circular segment (distance cut into round stock to make 6 flats).

Hexagon

$A = 2.598S^2 = 2.598R^2 = 3.464r^2$
$S = D \times .5 = R = 1.155r$
$d = D \times .067$
$r = .866S$

A = Area; S = Side; D = Diagonal; R = Radius of circumscribed circle; r = Radius of inscribed circle; d = Height of circular segment (distance cut into round stock to make 8 flats).

Octagon

$A = 4.828S^2$
$S = D \times .3827$
$R = 1.3075 = 1.082r$
$r = 1.207S$
$d = D \times .038$

A = Area; D = Diameter; R = Radius; C = Circumference

Circle

$A = \pi R^2 = 3.1416R^2 = .7854D^2$
$C = 2\pi R = 6.2832R = 3.1416D$
$R = C \div 6.2832$

A = Area; l = Length of arc; C = Chord length; R = Radius; λ = angle (in degrees); d = Height of circular segment (distance cut into round stock to produce a flat of a given width).

Cicular segment

$A = \frac{1}{2}[Rl - C(R - d)]$
$C = 2\sqrt{d(2R - d)}$
$d = R - \frac{1}{2}\sqrt{4R^2 - C^2}$
$l = .01745R\lambda$
$\lambda = \dfrac{57.296l}{R}$

Note: To find the weight of a metal bar of any cross-sectional shape, find the area of cross section, multiply by the length of the bar in inches and by the weight in pounds per cubic inch of the material.

The weights of some metals in pounds per cubic inch are as follows: steel, .284; aluminum, .0975; bronze, .317; copper, .321; lead, .409; and silver, .376.

Table 6

Allowances for Fits of Bores in Inches

Diameter (in.)	Running and Sliding Fits (free to rotate and free to slide)	Standard Fits (readily assembled)	Driving Fits (permanent assembly, light drive)	Forced Fits (permanent assembly with hydraulic press)
Up to ½	+.0005 to +.001	+.00025 to +.0005	−.0005	−.00075
½ to 1	+.001 to +.0015	+.0003 to +.001	−.0075	−.0015
1 to 2	+.0015 to +.0025	+.0004 to +.0015	−.001	−.0025
2 to 3½	+.002 to +.003	+.0005 to +.002	−.0015	−.0035
3½ to 6	+.003 to +.004	+.00075 to +.003	−.002	−.0045

Note: In this table, shaft sizes are considered as nominal and bore sizes are varied for fits, thus a negative fit is an interference (press) fit and a positive fit is a free or loose fit.

Table 7

Tapers and Corresponding Angles

Taper per Foot	Included Angle		Angle with Centerline		Taper per Inch
	Degrees	Minutes	Degrees	Minutes	
1/8	0	36	0	18	.0104
3/16	0	54	0	27	.0156
1/4	1	12	0	36	.0208
5/16	1	30	0	45	.0260
3/8	1	47	0	53	.0313
7/16	2	5	1	2	.0365
1/2	2	23	1	11	.0417
9/16	2	42	1	21	.0469
5/8	3	00	1	30	.0521
11/16	3	18	1	39	.0573
3/4	3	35	1	48	.0625
13/16	3	52	1	56	.0677
7/8	4	12	2	6	.0729
15/16	4	28	2	14	.0781
1	4	45	2	23	.0833
1 1/4	5	58	2	59	.1042
1 1/2	7	8	3	34	.1250
1 3/4	8	20	4	10	.1458
2	9	32	4	46	.1667
2 1/2	11	54	5	57	.2083
3	14	16	7	8	.2500
3 1/2	16	36	8	18	.2917
4	18	56	9	28	.3333
4 1/2	21	14	10	37	.3750
5	23	32	11	46	.4167
6	28	4	14	2	.5000

Source: Warren T. White, John E. Neely, Richard R. Kibbe, and Roland O. Meyer, *Machine Tools and Machining Practices*, John Wiley and Sons, Inc., Copyright © 1977, New York.